HANDBOOK OF DIGITAL LOGIC. . . .

With Practical Applications

SAM COWAN

PRENTICE-HALL, INC.
ENGLEWOOD CLIFFS, NEW JERSEY

(Business and Professional Division)

Prentice-Hall International, Inc., *London*
Prentice-Hall of Australia, Pty. Ltd., *Sydney*
Prentice-Hall of Canada, Inc., *Toronto*
Prentice-Hall of India Private Ltd., *New Delhi*
Prentice-Hall of Japan, Inc., *Tokyo*
Prentice-Hall of Southeast Asia Pte. Ltd., *Singapore*
Whitehall Books, Ltd., Wellington, *New Zealand*
Editora Prentice-Hall do Brasil Ltda., *Rio de Janeiro*
Prentice-Hall Hispanoamericana, *S.A., Mexico*

PRENTICE-HALL, INC.
Englewood Cliffs, N. J.

First Printing

Editor: George E. Parker

Library of Congress Cataloging in Publication Data

Cowan, Sam
Handbook of digital logic . . . with practical applications.

Includes index.
1. Digital electronics. 2. Logic circuits.
I. Title.
TK7868.D5C68 1985 621.3815 84-16049

ISBN 0-13-377193-8

Printed in the United States of America

How This Book Will Help You Accomplish More with Digital Logic Circuits. . . .

This handbook contains all of the information the electronics professional needs in order to understand and apply digital logic. It covers the entire field from Boolean algebra and logic circuits, to microprocessors and memory chips. It does so in a straightforward, practical manner with worked-out examples each step of the way.

The entire field of digital logic is covered in this one handbook, from number systems and Boolean algebra, to encoding schemes for magnetic recording, to circuits used to implement common error correction codes. This broad range of essential information will provide a handy, one-stop reference source. Refer to it whenever you need to simplify a logic circuit, use a Hamming code, or look up the instruction codes for a common microprocessor.

Consider the following examples:

A. The chapter on microprocessors has a step-by-step example of the functioning of a microprocessor.

B. Chapter 7 explains the memory devices used on computer systems.

C. Chapter 6 has an example of a complete working microprocessor system.

D. Charts for rapid conversion between number systems are given in Chapter 1.

E. The chapter on Boolean algebra shows how to simplify logic circuits by using Karnaugh maps or by using algebra.

F. The progression of flip-flops from simple latches to master-slave JK flip-flops and counters is shown in Chapter 5.

G. Chapter 4 is full of working logic circuits covering arithmetic circuits and circuits used to convert between different number systems.

H. Chapter 12 contains all the data necessary to use the 74XX/54XX series of T^2L logic chips.

Most current books on computers and digital logic limit their discussion of error codes to simple parity checks. While parity checks are useful and important, they are only the beginning of a much larger field of error detection and correction coding. Most 16K and 64K ram circuits use a Hamming or modified Hamming code to detect and correct errors. Disc and tape storage devices use binary cyclic codes to detect and correct errors. Each time the storage capacity of memory devices increases, the importance of error correction and detection also increases. This handbook gives *practical* working examples of circuits used to implement these codes.

The utility value of this handbook is illustrated by the following examples:

A. A step-by-step example of the action of a (7,4) Hamming code when a one-bit error occurs.

B. Two different ways to build counters to count in any base system.

C. Examples of CMOS, ECL, I^2L and other logic families.

D. Step-by-step illustrations of NRZI and MFM data encoding modes.

E. A complete list of the instruction codes for the 8080 and 6800 microprocessors.

Several worked-out examples, showing how to simplify a logic circuit and use a minimum number of gates, are given in Chapter 2. Sometimes the objective is not to use a minimum number of gates but to be able to use the types of gates that are available. These examples are also worked out step by step.

Chapter 3 shows specific examples of all the common logic families. Comparisons of the different types are given in terms of speed, power consumption, and compatibility.

Also included in this handbook is a chapter covering all the popular T^2L logic chips. The schematics, pin assignments, and other useful data are supplied, so the reader doesn't have to use other reference sources.

This handbook will be a great aid to all electronics professionals active in the digital field. It contains, in one reference source, all the information necessary to save you time and money.

Sam Cowan

**This book is dedicated to my parents
Byron and Edna Cowan.**

ACKNOWLEDGMENT

I would like to thank my wife Kathy for her efforts in typing the manuscript, correcting my spelling, and allowing me the time to write.

CONTENTS

1

Using Number Systems

INTRODUCTION

The decimal number system was developed long before computers and digital systems existed. It is not surprising then that other number systems are more applicable to digital electronics than the decimal system. Most digital systems use binary, octal, hexadecimal, and other schemes related to a power of two. Because of this it is necessary to know how to perform arithmetic and other operations using these number systems. To help understand these number systems, it is necessary to understand the basics of the decimal system.

UNDERSTANDING THE DECIMAL SYSTEM

The decimal or base ten system gets its name from the fact that it uses ten unique symbols. These symbols are:

0, 1, 2, 3, 4, 5, 6, 7, 8, 9.

From these symbols it is possible to represent any quantity. This is done by considering two characteristics:

1. The symbol itself.
2. The position of the symbol within the number.

The position of the symbol within the number is referred to as "positional notation," and it is this position that gives a symbol its weight. The symbol can represent ones, tens, hundreds, or even millions, depending on its position within the number.

Example

The number 437 is read "four hundred and thirty-seven." The 4, because of its position, represents hundreds. The 3 represents tens and the 7 represents ones, because of their positions within the number.

In the decimal system, positions are weighted by powers-of-ten. The first digit on the right represents ones or 10^0. The next digit to the left represents tens or 10^1. Each successive position to the left represents a power of ten one higher than its neighbor to the right.

Example

Consider the number 1253. The positions are weighted as follows:

3 ones or 10^0	=	3
5 tens or 10^1	=	50
2 hundreds or 10^2	=	200
1 thousands or 10^3	=	1000
Total	=	1253

It is also important to understand the rules involved when counting in the decimal system (see Figure 1-1). When all the symbols have been used in one column, the next column is started or indexed.

With the decimal system, when you count through nine, you must start a new column to the left, i.e., 10. When you reach 19, you must index the column to the left, or 20. When

1	6	11	16
2	7	12	17
3	8	13	18
4	9	14	19
5	(10)	15	(20)

Figure 1-1

99 is reached, both columns are full so a new column must be started. The next number is 100.

When the sequence of symbols is exhausted in one column, the column to the left is started or indexed.

These three rules apply to all number systems:

1. The number of unique symbols used determines the "base" of the system. It is possible to construct number systems of any base that is two or greater.
2. In positional notation, a digit is weighted by its position within the number. The weight of each position relates to a power of the base system. In general, the positions have weight thus:

   $X^n + \ldots X^3 + X^2 + X^1 + X^0$
3. When counting in base X, the first column goes from 0 through X-1. At this point the column to the left must be started or indexed.

EXAMPLES OF OTHER BASE SYSTEMS

To illustrate these rules on another base system, let's use for an example base five. By definition there must be five symbols in this system:

0, 1, 2, 3, 4.

Counting in base five would be as seen in Figure 1-2. When all the symbols have been used in the right-hand column, the column to the left is started or indexed.

0	12	24	41
1	13	30	42
2	14	31	43
3	20	32	44
4	21	33	100
10	22	34	
11	23	40	

Figure 1-2

The positional notation in base five has the digits weighted in powers of five. The first column is $5^0 = 1$, the second is $5^1 = 5$, the third $5^2 = 25$, etc.

Example

In the number 423 base 5, there are

4 twenty-fives	5^2
2 fives	5^1
3 ones	5^0

A number system in common use in digital systems is octal or base eight. The symbols for this system are:

0, 1, 2, 3, 4, 5, 6, 7.

Counting in this system is shown in Figure 1-3.

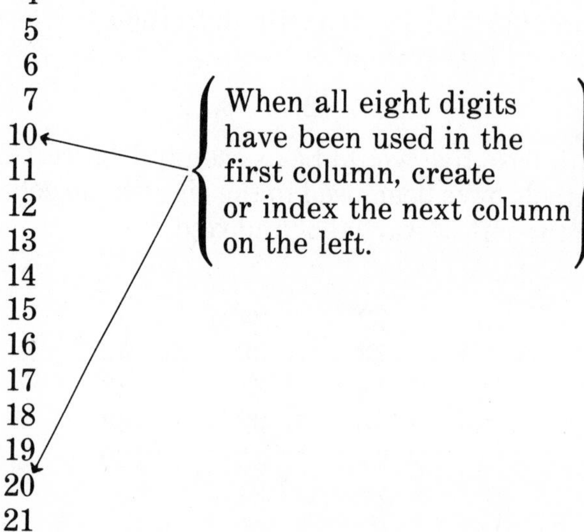

Figure 1-3

The positional notation for base eight has the digits weighted in powers of eight. The first column is $8^0 = 1$, the second is $8^1 = 8$, and the third is $8^2 = 64$.

Example

In the octal number 615 base 8, there are

6 sixty-fours	8^2
1 eights	8^1
5 ones	8^0

The examples so far have shown number systems with bases less than ten. The symbols used in these systems are the same as those used in the decimal system. It is when using a number system with a base larger than ten that it is necessary to create some new symbols. It is common practice to use the letters of the alphabet as symbols in number systems. One good example of this is base sixteen or hexadecimal. The sixteen symbols used in hexadecimal are:

0, 1, 2, 3, 4, 5, 6, 7, 8, 9, A, B, C, D, E, F.

For counting in base 16, see Figure 1-4.

The positional notations for base sixteen have digits weighted in powers of sixteen. The first column is $16^0 = 1$, the second $16^1 = 16$, the third is $16^2 = 256$, etc.

Example

In the hexadecimal number 3A7 base 16, there are

3 two-hundred-fifty-sixes	16^2
(A or 10) sixteens	16^1
7 ones	16^0

1	6	B	10	15	1A	1F
2	7	C	11	16	1B	20
3	8	D	12	17	1C	21
4	9	E	13	18	1D	22
5	A	F	14	19	1E	23

Figure 1-4

Using the same basic rules, number systems may be generated using any base except zero and one. The number systems most commonly used in digital electronics are binary (base 2), octal (base 8), and hexadecimal (base 16). Binary is used because its two symbols can easily be represented by two different voltage levels. Octal and hexadecimal are used because they are base systems that are powers of two, and they can be used by the proper grouping of binary digits.

HOW TO CONVERT BETWEEN SYSTEMS

Converting between base systems is done by three basic types of conversion:

1. Converting a number from base X to base 10.
2. Converting a number from base 10 to base X.
3. Converting between number systems that are powers of 2; i.e., binary, octal, hexadecimal.

With the first two methods it is possible to convert between any two base systems by first converting to base 10 and then converting back.

Converting from Base X to Base 10

Conversion from any base number to base 10 can be done by following these three steps:

1. Write down the weight of each position.
2. Multiply the weight by the digit in that position.
3. Add the results.

Example 1:

Convert 321 base 6 to base 10.
The positional weights in base 6 are:

$$
\begin{array}{lcrcr}
6^0 = 1 & \longrightarrow & 1 \times 1 & = & 1 \\
6^1 = 6 & \longrightarrow & 2 \times 6 & = & 12 \\
6^2 = 36 & \longrightarrow & 3 \times 36 & = & 108 \\
\hline
 & & \text{Total} & & 121
\end{array}
$$

321 base 6 equals 121 base 10.

Example 2:

Convert ABC base 16 to base 10.
The positional weights in base 16 are:

$$
\begin{array}{lcr}
16^0 = 1 & \longrightarrow & 1 \times C = 12 \\
16^1 = 16 & \longrightarrow & 16 \times B = 176 \\
16^2 = 256 & \longrightarrow & 256 \times A = 2560 \\
\hline
 & & \text{Total} \quad 2748
\end{array}
$$

ABC base 16 equals 2748 base 10.

Example 3:

Convert 11010 base 2 to base 10.
The positional weights in base 2 are:

$$
\begin{array}{lcr}
2^0 = 1 & \longrightarrow & 1 \times 0 = 0 \\
2^1 = 2 & \longrightarrow & 2 \times 1 = 2 \\
2^2 = 4 & \longrightarrow & 4 \times 0 = 0 \\
2^3 = 8 & \longrightarrow & 8 \times 1 = 8 \\
2^4 = 16 & \longrightarrow & 16 \times 1 = 16 \\
\hline
 & & \text{Total} \quad 26
\end{array}
$$

11010 base 2 equals 26 base 10.

Converting from binary (base 2) to decimal (base 10) and back the other way is common in digital electronics. A table for doing this is given later in this chapter. To convert by the methods given above, it is necessary to know the positive and negative powers of two. Figure 1-5 gives these values for powers of two up to twenty.

Converting from Base 10 to Base X

Converting decimal numbers to any given base system can be done by the following steps:

1. Write down the positional weights of the base system to be converted to.
2. Determine the largest weight necessary to convert. (This is the first weight less than the decimal number.)
3. Sequentially divide each weight into the number until there is no remainder.

N	2^N	2^{-N}
0	1	1.0
1	2	0.50
2	4	0.25
3	8	0.125
4	16	0.0625
5	32	0.03125
6	64	0.015625
7	128	0.0078125
8	256	0.00390625
9	512	0.001953125
10	1024	0.0009765625
11	2048	0.00048828125
12	4096	0.000244140625
13	8192	0.0001220703125
14	16382	0.00006103515625
15	32768	0.000030517578125
16	65536	0.0000152587890625
17	131072	0.00000762939453125
18	262144	0.000003814697265625
19	524288	0.0000019073486328125
20	1048576	0.00000095367431640625

Figure 1-5

Example 1:

Convert 58 base 10 to base 8 (octal).

1. The positional weights in octal are:

$$8^0 = 1$$
$$8^1 = 8$$
$$8^2 = 64$$

2. The largest weight necessary to convert is $8^1 = 8$, since 64 is larger than 58.

3. $\left\{\begin{array}{l}\text{Divide 8 into 58.}\\ 58 \div 8 = 7^+,\ 8 \times 7 = 56.\\ 58 - 56 = 2.\end{array}\right\} \longrightarrow \left\{\begin{array}{l}\text{The first}\\ \text{octal digit}\\ \text{is 7.}\end{array}\right.$

$\left\{\begin{array}{l}\text{Divide 1 into 2.}\\ 2 \div 1 = 2,\ 2 \times 1 = 2.\end{array}\right\} \longrightarrow \left\{\begin{array}{l}\text{The second octal}\\ \text{digit is 2.}\end{array}\right.$

58 base 10 equals 72 octal.

Example 2:

Convert 948 decimal to base 16.

1. The positional weights of base 16 are:

$$\begin{aligned}16^0 &= 1\\ 16^1 &= 16\\ 16^2 &= 256\\ 16^3 &= 4096\end{aligned}$$

2. The largest weight necessary to convert is $16^2 = 256$, since 4096 is larger than 948.

3. Divide 948 by 256.
$948 \div 256 = 3^+,\ 256 \times 3 = 768.$
$948 - 768 = 180.$
} The first hexadecimal digit is 3.

Divide 180 by 16.
$180 \div 16 = 11^+,\ 11 \times 16 = 176.$
$180 - 176 = 4.$
} The second hexadecimal digit is B.

Divide 4 by 1.
$4 \div 1 = 4,\ 4 \times 1 = 4.$
$4 - 4 = 0.$
} The third hexadecimal digit is 4.

948 decimal equals 3B4 hexadecimal.

Conversion between two base systems where neither base is decimal can be done by first converting to decimal and then to the desired base system.

Example:

Convert 32 base 5 to base 3.

$$\begin{aligned}3 \times 5^1 &= 15\\ 2 \times 5^0 &= \underline{\ \ 2}\\ &\ \ 17\end{aligned}$$

32 base 5 equals 17 base 10.
The positional weights of base 3 are:

$3^0 = 1$
$3^1 = 3$
$3^2 = 9$
$3^3 = 27$

$17 \div 9 = 1^+, 9 \times 1 = 9$
$17 - 9 = 8$ } First digit is 1.

$8 \div 3 = 2^+, 2 \times 3 = 6$
$8 - 6 = 2$ } Second digit is 2.

$2 \div 1 = 2, 2 \times 1 = 2$
$2 - 2 = 0$ } Third digit is 2.

32 base 5 equals 122 base 3.

If the conversion is between base systems that are powers of two, such as binary, octal, and hexadecimal, the conversion is simplified. It is only necessary to group the digits in the proper order.

Example 1:

Convert 10110011 binary to octal.
Since eight is two cubed, it is necessary to arrange the binary digits into groups of three:

10110011 (10) (110) (011)

Each group of three will become an octal digit.

(10) (110) (011)
↓ ↓ ↓
2 6 3

10110011 binary equals 263 octal.

Example 2:

Convert 9B3 hexadecimal to binary.
Since sixteen is two raised to the fourth power, each hexadecimal digit is converted to four binary digits.

9 = 9 = (1001)
B = 11 = (1011)
3 = 3 = (0011)

9B3 base 16 equals 100110110011 base 2.

USING DECIMALS IN DIFFERENT BASE SYSTEMS

To understand decimals in various number systems, it is necessary to understand them in base 10. To understand decimals in base 10, consider the number 79.158.

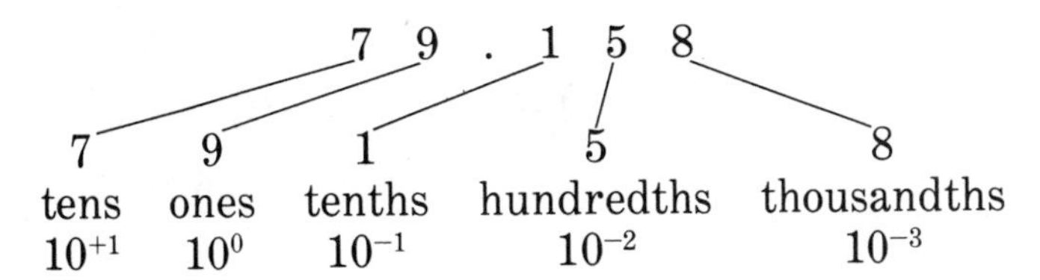

The decimal point is the dividing point between positive powers of ten and negative powers of ten. The digits to the right of the decimal point are weighted by negative powers of the base system. It is true that in any base system, the digits to the right of the decimal point are weighted by negative powers of the base. The binary number 101.110 is weighted as follows:

1 0 1 . 1 1 0

$(2^2 = 4\text{'s})$ $(2^1 = 2\text{'s})$ $(2^0 = 1\text{'s})$ $(2^{-1} = \frac{1}{2}\text{'s})$ $(2^{-2} = \frac{1}{4}\text{'s})$ $(2^{-3} = \frac{1}{8}\text{'s})$

In the same manner, the octal number 67.25 is weighted as:

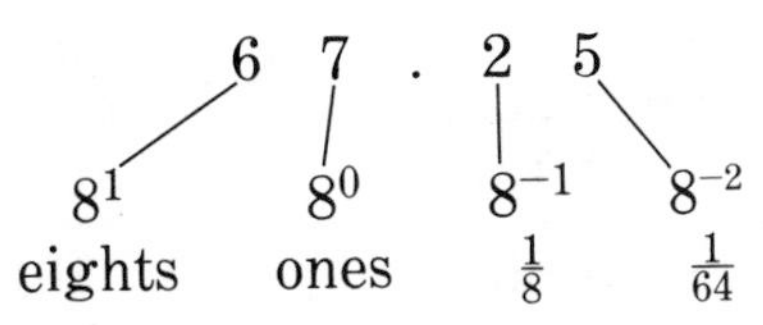

DOING ARITHMETIC IN ANY BASE SYSTEM

Addition, subtraction, multiplication, and division can be performed using any number system. The rules are the same as those used in base 10. The following are examples of arithmetic operations with some of the most commonly used number systems:

Decimal

ADDITION

```
  4391
+ 2106
------
  6497
```

SUBTRACTION

```
   8 11
  43̸9̸1
- 2106
------
  2285
```

MULTIPLICATION

```
    53
×   27
------
   371
  106
------
  1431
```

DIVISION

```
   4.80
5 |24.00
   20
   --
    40
    40
    --
```

When adding and subtracting numbers in base 10, it is necessary to carry to the column on the left or borrow from that column. This happens when the sum is larger than the base (ten), or when subtracting and the lower digit is larger than the upper digit. The same rules apply when doing arithmetic with other number systems.

Binary

ADDITION

```
  10111   ⟶  23₁₀
+ 01101   ⟶  13₁₀
-------      ----
 100100   ⟶  36₁₀
```

SUBTRACTION

```
  10
  1̸0111  ⟶  23₁₀
- 01101  ⟶  13₁₀
-------     ----
   1010  ⟶  10₁₀
```

MULTIPLICATION

```
     1101 ⟶  13₁₀
   ×  111 ⟶   7₁₀
---------    ----
     1101
    1101
   1101
---------    ----
  1011011      91
```

DIVISION

```
    10.01
11 |111.00
    11
    ----
    0100
      11
    ----
      10 remainder
```

Octal

ADDITION

$$\begin{array}{rcr} \overset{1}{} 36_8 & \longrightarrow & 30_{10} \\ +\ 43_8 & \longrightarrow & 35_{10} \\ \hline 101_8 & \longrightarrow & 65_{10} \end{array}$$

SUBTRACTION

$$\begin{array}{rcr} \overset{3\ 13}{\not{4}3_8} & \longrightarrow & 35_{10} \\ -\ 36_8 & \longrightarrow & 30_{10} \\ \hline 5_8 & \longrightarrow & 5_{10} \end{array}$$

MULTIPLICATION

$$\begin{array}{rcr} 45_8 & \longrightarrow & 37_{10} \\ \times\ 32_8 & \longrightarrow & 26_{10} \\ \hline 112 & & 222 \\ 157\ \ & & 74\ \ \\ \hline 1702_8 & \longrightarrow & 962_{10} \end{array}$$

DIVISION

$$\begin{array}{r|l} & 2.66 \\ \hline 5 & 16.0 \\ & 12 \\ \hline & \ \ 40 \\ & \ \ 36 \\ \hline & \quad 40 \text{ remainder} \end{array}$$

Hexadecimal

ADDITION

$$\begin{array}{rcr} \overset{5}{\not{3}}B2 & \longrightarrow & 946_{10} \\ +\ A5 & \longrightarrow & 165_{10} \\ \hline 817 & \longrightarrow & 1111_{10} \end{array}$$

SUBTRACTION

$$\begin{array}{rcr} 3\overset{A\ 1}{\not{B}2} & \longrightarrow & 964_{10} \\ -\ A5 & \longrightarrow & 165_{10} \\ \hline 30D & \longrightarrow & 781_{10} \end{array}$$

MULTIPLICATION

$$\begin{array}{rcr} F3 & \longrightarrow & 243_{10} \\ \times\ 55 & \longrightarrow & 85_{10} \\ \hline 4BF & & 1215 \\ 4BF\ \ & & 1944\ \ \\ \hline 50AF & - & 20655_{10} \end{array}$$

DIVISION

$$\begin{array}{r|l} & 12.19 \\ \hline A & B5.00 \\ & A \\ \hline & 15 \\ & 14 \\ \hline & \ \ 10 \\ & \ \ \ A \\ \hline & \ \ \ 60 \\ & \ \ \ 5A \\ \hline & \quad 6 \text{ remainder} \end{array}$$

EXAMPLES OF CODED NUMBERS

In digital electronics it is common to represent numbers in terms of a code. These codes were devised for a variety of reasons: some for error detection, some for ease of use in electronic circuits, and some to convert alphanumeric symbols into binary numbers. The most commonly used codes are listed in this section.

B.C.D.—Binary Coded Decimal or 8421 Code

The BCD code is used to represent decimal numbers in a convenient binary form. To write the decimal digits (zero through 9) in binary, it is necessary to use four binary digits. Three binary digits could only represent zero through seven, and five binary digits could represent zero through thirty-one.

In BCD each decimal digit is represented by a four-digit binary group.

Example 1:

Encode the decimal number 529 in B.C.D.

5 equals 0101
2 equals 0010
9 equals 1001

529 base 10 equals (0101) (0010) (1001) BCD.

Example 2:

Encode the decimal number 873 in BCD.

8 equals 1000
7 equals 0111
3 equals 0011

873 base 10 equals (1000) (0111) (0011) BCD.

Excess 3 Code

The excess 3 code is derived in the same way as the B.C.D. code except that before conversion into binary each digit is increased by *three*.

Example 1:

Encode the decimal number 893 in excess 3.

$8 + 3 = 11$ ———————— 1011
$9 + 3 = 12$ ———————— 1100
$3 + 3 = 6$ ———————— 0110

893 base 10 equals (1011) (1100) (0110) excess 3.

Example 2:

Encode the decimal number 123 in excess 3.

$1 + 3 = 4$ ———————— 0100
$2 + 3 = 5$ ———————— 0101
$3 + 3 = 6$ ———————— 0110

123 base 10 equals (0100) (0101) (0110) excess 3.

Gray Code

The gray code is a sequence of binary numbers in which one, and only one, digit changes in successive numbers. It is also called the "unit-distance" code and is used in Karnaugh maps, error detection schemes, and other places in digital electronics. (See Figure 1-6.)

DECIMAL	GRAY CODE
0	0000
1	0001
2	0011
3	0010
4	0110
5	0111
6	0101
7	0100
8	1100
9	1101
10	1111
11	1110
12	1010
13	1011
14	1001
15	1000

Figure 1-6

2-Out-of-5 Code

Used almost exclusively for error detection purposes, the 2-out-of-5 code consists of five binary digits for each decimal digit. Of these five digits, two are ones and three are zeros. Along with the fact that two and only two digits will be ones, the 2-out-of-5 code is an example of even parity. (See Figure 1-7.)

DECIMAL	2-OUT-OF-5
0	00011
1	00101
2	00110
3	01001
4	01010
5	01100
6	10001
7	10010
8	10100
9	11000

Figure 1-7

USING COMPLEMENTS

Complements are a way of representing numbers in such a way that subtraction can be performed by doing addition. In a digital circuit this allows addition and subtraction to be done by the same circuit.

Complements can be used in any base system, and they are generated and used according to the same basic rules.

Decimal (Base 10) Complements

The 10's complement is formed by subtracting each digit from the number 9 and adding 1 to the least significant digit. Figure 1-8 gives some decimal numbers along with their 10's complement.

Subtraction is performed by adding the complement as shown in the following examples:

Example 1:

Using complements, subtract 25 from 93.

Number	10's complement
93	07
27	73
158	842
32	68

Figure 1-8

Normal subtraction

```
 93
-25
---
 68
```

10's complement

```
   93
+  75 ⟶ 10's complement of 25
------
  1̸68
```

The 1 carry is dropped.

Example 2:

Using complements, subtract 93 from 200.

Normal subtraction

```
 200
-  93
-----
 107
```

10's complement

```
  200
+  07   10's complement of 93
------
  2̸07
  107
```

1 is dropped from the carry.

The 9's complement is formed by subtracting each digit from 9. It is unlike the 10's complement since a 1 is not added to the least significant digit. This addition is done when subtracting. Figure 1-9 shows the 9's complement for some random decimal numbers.

Subtraction is performed by adding the 9's complement and then adding the 1 carry to the result.

Example:

Using the 9's complement, subtract 44 from 59.

Normal subtraction	9's complement
59	59
− 44	+ 55
15	114
	+ 1
	15—the carry is added.

Decimal number	9's complement
22	77
87	12
45	54
67	32
39	60

Figure 1-9

Binary (Base 2) Complements

Similar to the complements in base ten, binary complements can be used to perform subtraction by adding numbers. Also similar to base ten, binary numbers have two complements: the 2's complement and the 1's complement.

The 2's Complement

The 2's complement is formed by inverting each digit in the binary number and then adding 1 to the least significant digit. The 2's complement for some random numbers is shown in Figure 1-10.

Subtraction in binary is accomplished by adding the 2's complement.

Example 1:

Using 2's complement, subtract 101011 from 111011.

Binary number	2's complement
1011	0101
1100	0100
1001	0111
0111	1011

Figure 1-10

Normal subtraction	2's complement
111011	111011
− 101011	+ 010101 2's complement
10000	~~1~~010000
	The carry is dropped.

Example 2:

Using the 2's complement, subtract 100101 from 101011.

Normal subtraction	2's complement
101011	101011
− 100101	+ 011011 2's complement
000110	~~1~~000110
	The carry is dropped.

The 1's complement is formed by inverting each digit of the binary number. The 1's complement for some random numbers is shown in Figure 1-11.

Binary number	1's complement
1011	0100
1100	0011
1001	0110
0111	1000

Figure 1-11

Subtraction is performed in binary by adding the 1's complement and then adding the carry.

Example:

Using the 1's complement, subtract 101011 from 111011.

Normal subtraction	1's complement
111011	111011
− 101011	+ 010100 1's complement
10000	1001111
	+ 1
	10000

BINARY TO DECIMAL CONVERSION CHART

Methods used to convert between different base systems are given in earlier sections of the chapter. These methods will work for any base system; however, it is sometimes easier and faster to look up the conversion in a chart. Figure 1-12 gives binary to decimal conversions for decimal numbers up to 0000. An example of how the chart is used is given in Figure 1-13.

HEX TO DECIMAL CONVERSION CHART

Figure 1-14 contains a hexadecimal to decimal conversion chart for decimal numbers up to 1023. An example of how the chart is used is given in Figure 1-15.

	0	1	2	3	4	5	6	7	8	9
0	0	1	10	11	100	101	110	111	1000	1001
1	1010	1011	1100	1101	1110	1111	10000	10001	10010	10011
2	10100	10101	10110	10111	11000	11001	11010	11011	11100	11101
3	11110	11111	100000	100001	100010	100011	100100	100101	100110	100111
4	101000	101001	101010	101011	101100	101101	101110	101111	110000	110001
5	110010	110011	110100	110101	110110	110111	111000	111001	111010	111011
6	111100	111101	111110	111111	1000000	1000001	1000010	1000011	1000100	1000101
7	1000110	1000111	1001000	1001001	1001010	1001011	1001100	1001101	1001110	1001111
8	1010000	1010001	1010010	1010011	1010100	1010101	1010110	1010111	1011000	1011001
9	1011010	1011011	1011100	1011101	1011110	1011111	1100000	1100001	1100010	1100011
10	1100100	1100101	1100110	1100111	1101000	1101001	1101010	1101011	1101100	1101101
11	1101110	1101111	1110000	1110001	1110010	1110011	1110100	1110101	1110110	1110111
12	1111000	1111001	1111010	1111011	1111100	1111101	1111110	1111111	10000000	10000001
13	10000010	10000011	10000100	10000101	10000110	10000111	10001000	10001001	10001010	10001011
14	10001100	10001101	10001110	10001111	10010000	10010001	10010010	10010011	10010100	10010101
15	10010110	10010111	10011000	10011001	10011010	10011011	10011100	10011101	10011110	10011111
16	10100000	10100001	10100010	10100011	10100100	10100101	10100110	10100111	10101000	10101001
17	10101010	10101011	10101100	10101101	10101110	10101111	10110000	10110001	10110010	10110011
18	10110100	10110101	10110110	10110111	10111000	10111001	10111010	10111011	10111100	10111101
19	10111110	10111111	11000000	11000001	11000010	11000011	11000100	11000101	11000110	11000111
20	11001000	11001001	11001010	11001011	11001100	11001101	11001110	11001111	11010000	11010001
21	11010010	11010011	11010100	11010101	11010110	11010111	11011000	11011001	11011010	11011011
22	11011100	11011101	11011110	11011111	11100000	11100001	11100010	11100011	11100100	11100101
23	11100110	11100111	11101000	11101001	11101010	11101011	11101100	11101101	11101110	11101111
24	11110000	11110001	11110010	11110011	11110100	11110101	11110110	11110111	11111000	11111001
25	11111010	11111011	11111100	11111101	11111110	11111111				

Figure 1-12

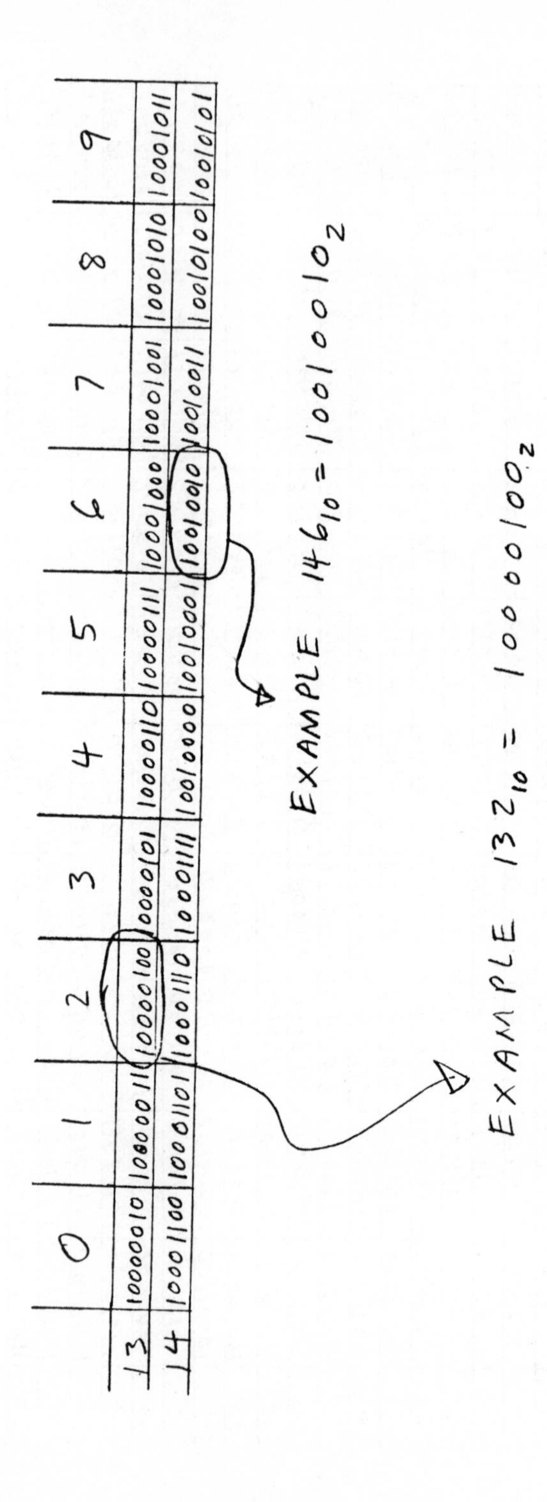

0
1
2
3
4
5
6
7
8
9
13
10000010
10000011
10000100
10000101
10000110
10000111
10001000
10001001
10001010
10001011
14
10001100
10001101
10001110
10001111
10010000
10010001
10010010
10010011
10010100
10010101
EXAMPLE $146_{10} = 10010010_2$
EXAMPLE $132_{10} = 10000100_2$

Figure 1-13

N	0	1	2	3	4	5	6	7	8	9	A	B	C	D	E	F
0	0000	0001	0002	0003	0004	0005	0006	0007	0008	0009	0010	0011	0012	0013	0014	0015
1	0016	0017	0018	0019	0020	0021	0022	0023	0024	0025	0026	0027	0028	0029	0030	0031
2	0032	0033	0034	0035	0036	0037	0038	0039	0040	0041	0042	0043	0044	0045	0046	0047
3	0048	0049	0050	0051	0052	0053	0054	0055	0056	0057	0058	0059	0060	0061	0062	0063
4	0064	0065	0066	0067	0068	0069	0070	0071	0072	0073	0074	0075	0076	0077	0078	0079
5	0080	0081	0082	0083	0084	0085	0086	0087	0088	0089	0090	0091	0092	0093	0094	0095
6	0096	0097	0098	0099	0100	0101	0102	0103	0104	0105	0106	0107	0108	0109	0110	0111
7	0112	0113	0114	0115	0116	0117	0118	0119	0120	0121	0122	0123	0124	0125	0126	0127
8	0128	0129	0130	0131	0132	0133	0134	0135	0136	0137	0138	0139	0140	0141	0142	0143
9	0144	0145	0146	0147	0148	0149	0150	0151	0152	0153	0154	0155	0156	0157	5018	0159
A	0160	0161	0162	0163	0164	0165	0166	0167	0168	0169	0170	0171	0172	0173	0174	0175
B	0176	0177	0178	0179	0180	0181	0182	0183	0184	0185	0186	0187	0188	0189	0190	0191
C	0192	0193	0194	0195	0196	0197	0198	0199	0200	0201	0202	0203	0204	0205	0206	0207
D	0208	0209	0210	0211	0212	0213	0214	0215	0216	0217	0218	0219	0220	0221	0222	0223
E	0224	0225	0226	0227	0228	0229	0230	0231	0232	0233	0234	0235	0236	0237	0238	0239
F	0240	0241	0242	0243	0244	0245	0246	0247	0248	0249	0250	0251	0252	0253	0254	0255
10	0256	0257	0258	0259	0260	0261	0262	0263	0264	0265	0266	0267	0268	0269	0270	0271
11	0272	0273	0274	0275	0276	0277	0278	0279	0280	0281	0282	0283	0284	0285	0286	0287
12	0288	0289	0290	0291	0292	0293	0294	0295	0296	0297	0298	0299	0300	0301	0302	0303
13	0304	0305	0306	0307	0308	0309	0310	0311	0312	0313	0314	0315	0316	0317	0318	0319
14	0320	0321	0322	0323	0324	0325	0326	0327	0328	0329	0330	0331	0332	0333	0334	0335
15	0336	0337	0338	0339	0340	0341	0342	0343	0344	0345	0346	0347	0348	0349	0350	0351
16	0352	0353	0354	0355	0356	0357	0358	0359	0360	0361	0362	0363	0364	0365	0366	0367
17	0368	0369	0370	0371	0372	0373	0374	0375	0376	0377	0378	0378	0380	0381	0382	0383
18	0384	0385	0386	0387	0388	0389	0390	0391	0392	0393	0394	0395	0396	0397	0398	0399
19	0400	0401	0402	0403	0404	0405	0406	0407	0408	0409	0410	0411	0412	0413	0414	0415
1A	0416	0417	0418	0419	0420	0421	0422	0423	0424	0425	0426	0427	0428	0429	0430	0431
1B	0432	0433	0434	0435	0436	0437	0438	0439	0440	0441	0442	0443	0444	0445	0446	0447
1C	0448	0449	0450	0451	0452	0453	0454	0455	0456	0457	0458	0459	0460	0461	0462	0463
1D	0464	0465	0466	0467	0468	0469	0470	0471	0472	0473	0474	0475	0476	0477	0478	0479
1E	0480	0481	0482	0483	0484	0485	0486	0487	0488	0489	0490	0491	0492	0493	0494	0495
1F	0496	0497	0498	0499	0500	0501	0502	0503	0504	0505	0506	0507	0508	0509	0510	0511

Figure 1-14

N	0	1	2	3	4	5	6	7	8	9	A	B	C	D	E	F
20	0512	0513	0514	0515	0516	0517	0518	0519	0520	0521	0522	0523	5024	0525	0526	0527
21	0528	0529	0530	0531	0532	0533	0534	0535	0536	0537	0538	0539	0540	0541	0542	0543
22	0544	0545	0546	0547	0548	0549	0550	0551	0552	0553	0554	0555	0556	0557	0558	0559
23	0560	0561	0562	0563	0564	0565	0566	0567	0568	0569	0570	0571	0572	0573	0574	0575
24	0576	0577	0578	0579	0580	0581	0582	0583	0584	0585	0586	0587	0588	0589	0590	0591
25	0592	0593	0594	0595	0596	0597	0598	0599	0600	0601	0602	0603	0604	0605	0606	0607
26	0608	0609	0610	0611	0612	0613	0614	0615	0616	0617	0618	0619	0620	0621	0622	0623
27	0624	0625	0626	0627	0628	0629	0630	0631	0632	0633	0634	0635	0636	0637	0638	0639
28	0640	0641	0642	0643	0644	0645	0646	0647	0648	0649	0650	0651	0652	0653	0654	0655
29	0656	0657	0658	0659	0660	0661	0662	0663	0664	0665	0666	0667	0668	0669	0670	0671
2A	0672	0673	0674	0675	0676	0677	0678	0679	0680	0681	0682	0683	0684	0685	0686	0687
2B	0688	0689	0690	0691	0692	0693	0694	0695	0696	0697	0698	0699	0700	0701	0702	0703
2C	0704	0705	0706	0707	0708	0709	0710	0711	0712	0713	0714	0715	0716	0717	0718	0719
2D	0720	0721	0722	0723	0724	0725	0726	0727	0728	0729	0730	0731	0732	0733	0734	0735
2E	0736	0737	0738	0739	0740	0741	0742	0743	0744	0745	0746	0747	0748	0749	0750	0751
2F	0752	0753	0754	0755	0756	0757	0758	0759	0760	0761	0762	0763	0764	0765	0766	0767
30	0768	0769	0770	0771	0772	0773	0774	0775	0776	0777	0778	0779	0780	0781	0782	0783
31	0784	0785	0786	0787	0788	0789	0790	0791	0792	0793	0794	0795	0796	0797	0798	0799
32	0800	0801	8082	0803	0804	0805	0806	0807	0808	0809	0810	0811	0812	0813	0814	0815
33	0816	0817	0818	0819	0820	0821	0822	0823	0824	0825	0826	0827	0828	0829	0830	0831
34	0832	0833	0834	0835	0836	0837	0838	0839	0840	0841	0842	0843	0844	0845	0846	0847
35	0848	0849	0850	0851	0852	0853	0854	0855	0856	0857	0858	0859	0860	0861	0862	0863
36	0864	0865	0866	0867	0868	0869	0870	0871	0872	0873	0874	0875	0876	0877	0878	0879
37	0880	0881	0882	0883	0884	0885	0886	0887	0888	0889	0890	0891	0892	0893	0894	0895
38	0896	0897	0898	0899	0900	0901	0902	0903	0904	0905	0906	0907	0908	0909	0910	0911
39	0912	0913	0914	0915	0916	0917	0918	0919	0920	0921	0922	0923	0924	0925	0926	0927
3A	0928	0929	0930	0931	0932	0933	0934	0935	0936	0937	0938	0939	0940	0941	0942	0943
3B	0944	0945	0946	0947	0948	0949	0950	0951	0952	0953	0954	0955	0956	0957	0958	0959
3C	0960	0961	0962	0963	0964	0965	0966	0967	0968	0969	0970	0971	0972	0973	0974	0975
3D	0976	0977	0978	0979	0980	0981	0982	0983	0984	0985	0986	0987	0988	0989	0990	0991
3E	0992	0993	0994	0995	0996	0997	0998	0999	1000	1001	1002	1003	1004	1005	1006	1007
3F	1008	1009	1010	1011	1012	1013	1014	1015	1016	1017	1018	1019	1020	1021	1022	1023

Figure 1-14 (continued)

N	0	1	2	3	4	5	6	7	8	9	A	B	C	D	E	F
14	0320	0321	0322	0323	0324	0325	0326	0327	0328	0329	0330	0331	0332	0333	0334	0335
15	0336	0337	0338	0339	0340	0341	0342	0343	0344	0345	0346	0347	0348	0349	0350	0351

156 hex = 342 decimal

142 hex = 322 decimal

Figure 1-15

2

Understanding and Applying Boolean Algebra

INTRODUCTION

Boolean Algebra is the mathematical theory on which all modern computers operate, from the large scientific computers to the small logic boards designed for a single purpose. A working knowledge of its concepts is essential for an understanding of digital logic.

This chapter outlines the basic concepts and gives worked-out examples wherever possible. The chapter begins with the three basic gates used in Boolean algebra and defines their input/output relationships. Since all logic circuits are made from these basic gates, the input/output relationships of complex logic circuits can be determined from knowledge of AND gates, OR gates and inverters.

A Boolean Algebra function can be expressed in one of three different ways:

1. truth table
2. switching function
3. logic diagram

Any one of the three contains all the information necessary to completely define the function and, given any one of the three, the other two can be derived. This chapter gives graphic, algebraic, and tabulation methods for deriving the desired form and for simplifying functions. There are practical problems at the end of the chapter that show how to derive a desired logic circuit.

THE THREE BASIC GATES

Boolean Algebra consists of three basic connectives, logical addition (the OR gate), logical multiplication (the AND gate), and negation (the inverter). All switching functions represent some combination of these three basic functions.

It is fundamental in Boolean Algebra that a variable may have only two values. These values are usually represented by zero (0) and one (1). They may represent a switch that is on or off, a relay that is energized or deenergized, or they may represent a voltage difference. In almost all practical circuits, ones and zeros are represented by a difference in voltage levels. The most common representation is +5 volts = (1), 0 volts = (0).

The "OR" Gate

The OR gate is defined as follows:

A. *Schematic*

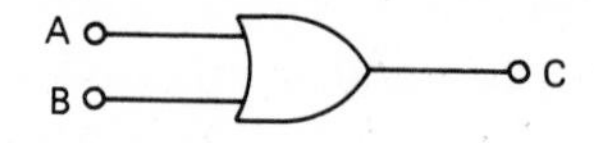

B. *Input/output*
The OR gate will have a 1 output when either of its inputs is a 1.

A	B	C
0	0	0
0	1	1
1	0	1
1	1	1

C. *Switching function*
The OR gate is known as logical addition and the (+) sign is used in the switching function.

$C = A + B$

The "AND" Gate

The AND gate is defined as follows:

A. *Schematic*

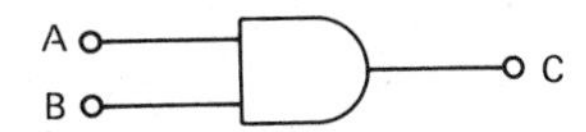

B. *Input/output*
The AND gate will have a 1 output when both inputs are 1.

A	B	C
0	0	0
0	1	0
1	0	0
1	1	1

C. *Switching function*
The AND gate is known as logical multiplication. The output is the product of the inputs.

$$C = AB$$

The Inverter

The inverter is defined as follows:

A. *Schematic*

B. *Input/output*
As the name implies, the output of the inverter is always the opposite of the input. This is also known as negation.

A	B
0	1
1	0

C. *Switching function*
The switching function is read B equals A "not" indicated by bar on top of A.

$$B = \overline{A}$$

INVERTED FUNCTIONS (NANDS, NORS)

It is common to use inverters on the outputs of AND/OR gates, so common in fact that these gates are given their own names. When the output of an AND gate is inverted, it is called a NAND gate and is drawn as shown in Figure 2-1. When the output of an OR gate is inverted, it is called a NOR gate and is drawn as shown in Figure 2-2.

Just as the small circle on the output indicates that the output has been inverted, it is common to use the same symbol on the input. For example, Figure 2-3 shows an inverted input on an AND gate.

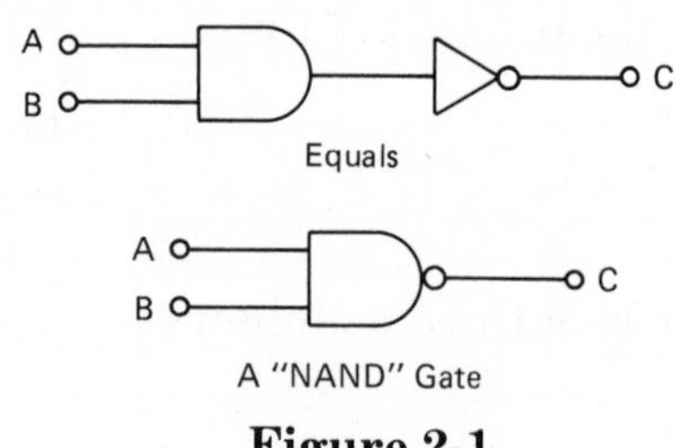

Figure 2-1

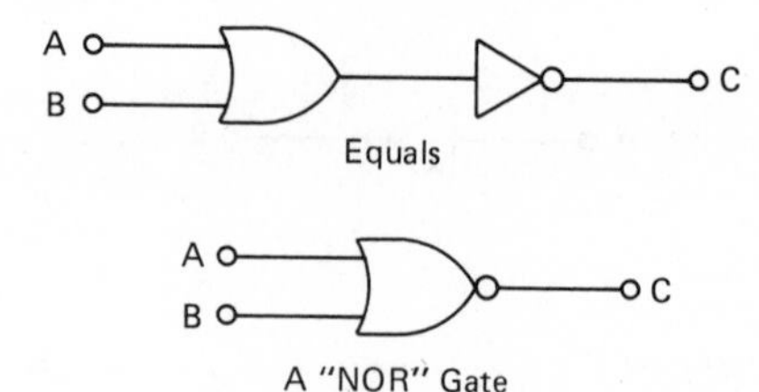

Figure 2-2

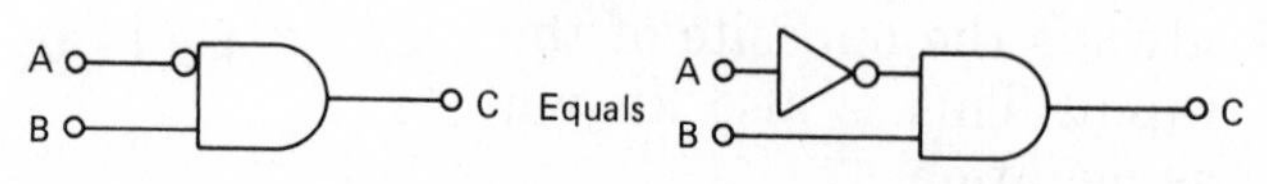

Figure 2-3

MULTIPLE INPUTS

The description of gates so far has been limited to gates with two inputs. The basic functions of AND and OR remain the same with gates that have more than two inputs. The AND gate must have all ones on the inputs to have a one out. The OR gate has a one out whenever any of inputs is a one. Truth Tables for three-input AND and OR gates are shown in Figure 2-4.

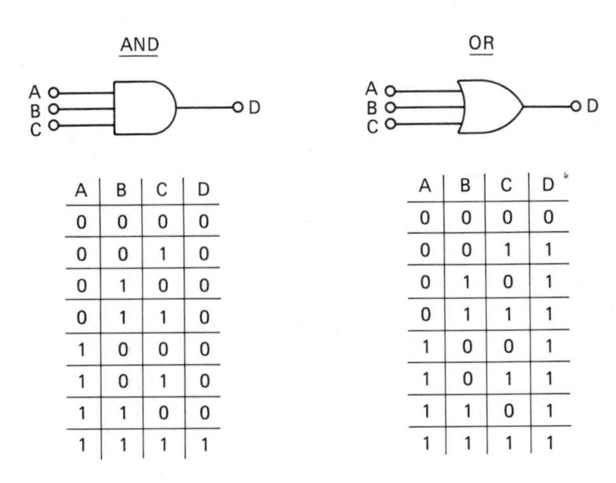

AND

A	B	C	D
0	0	0	0
0	0	1	0
0	1	0	0
0	1	1	0
1	0	0	0
1	0	1	0
1	1	0	0
1	1	1	1

OR

A	B	C	D
0	0	0	0
0	0	1	1
0	1	0	1
0	1	1	1
1	0	0	1
1	0	1	1
1	1	0	1
1	1	1	1

Figure 2-4

TRUTH TABLES

A Truth Table is a map showing the output of a logic circuit for each possible input. The inputs are usually arranged so that, from top to bottom, they are sequential binary numbers (see Figure 2-5). The three-input truth table has eight combinations of inputs. These are represented by the binary numbers 000 through 111. The four-input truth table has 16 combinations of inputs. These are represented by the binary numbers 0000 through 1111. Figure 2-5 shows a three-input table and a four-input table along with the decimal number equivalent of the binary input. In general, the total combinations of inputs can be determined from:

Input combinations = 2^N

Where: N = number of inputs

Example: A five-input table has $2^5 = 32$ possible combinations of inputs.

A	B	C	OUT
0	0	0	
0	0	1	
0	1	0	
0	1	1	
1	0	0	
1	0	1	
1	1	0	
1	1	1	

THREE-INPUT TABLE

DECIMAL	A	B	C	D	OUT
0	0	0	0	0	
1	0	0	0	1	
2	0	0	1	0	
3	0	0	1	1	
4	0	1	0	0	
5	0	1	0	1	
6	0	1	1	0	
7	0	1	1	1	
8	1	0	0	0	
9	1	0	0	1	
10	1	0	1	0	
11	1	0	1	1	
12	1	1	0	0	
13	1	1	0	1	
14	1	1	1	0	
15	1	1	1	1	

FOUR-INPUT TABLE

Figure 2-5

A truth table also has several different possible outputs. Consider the two-input truth table in Figure 2-6. There are $2^2 = 4$ input combinations 00, 01, 10, 11. Each possible input

X_1, X_2, X_3, X_4 CAN HAVE 16 DIFFERENT COMBINATIONS

X_1	X_2	X_3	X_4	
0	0	0	0	0
0	0	0	1	1
0	0	1	0	2
0	0	1	1	3
0	1	0	0	4
0	1	0	1	5
0	1	1	0	6
0	1	1	1	7
1	0	0	0	8
1	0	0	1	9
1	0	1	0	10
1	0	1	1	11
1	1	0	0	12
1	1	0	1	13
1	1	1	0	14
1	1	1	1	15

0–15 = 16
COMBINATIONS

A	B	OUT
0	0	X_1
0	1	X_2
1	0	X_3
1	1	X_4

A	B	OUT 1	OUT 2	OUT 3	OUT 4	OUT 5
0	0	0	0	0	0	0
0	1	0	0	0	0	1 →
1	0	0	0	1	1	0
1	1	0	1	0	1	0

$$\text{OUTPUTS} = (2)^{2^N}$$

FOR N = 2 $\quad 2^2 = 4 \quad 2^4 = 16$ COMBINATIONS

Figure 2-6

has a corresponding output. In the figure the outputs are given as X_1, X_2, X_3, X_4. Since there are four digits, there must be $2^4 = 16$ possible combinations of these digits. In general, the total combinations of outputs can be determined from:

Output combinations $= (2)^{2^N}$

Where: N = number of inputs

Example: A four-input table has $2^4 = 16$, $2^{16} = 65{,}536$ different possible combinations of outputs.

How to Generate Switching Functions from Truth Tables

There are three basic ways of expressing a Boolean Algebra function, a truth table, a switching function, and a logic diagram. Given any of the three, the other two can be determined.

Figures 2-7 and 2-8 show how the switching functions relate to the logic diagram. The AND gate implies multiplication, the OR gate implies addition, and the inverter is shown in the switching function by the (—) sign above the variable. When multiple gate networks are used, the switching function can be found by proceeding through the network from input to the output and writing the switching function at each point. This is shown in Figure 2-8 C.

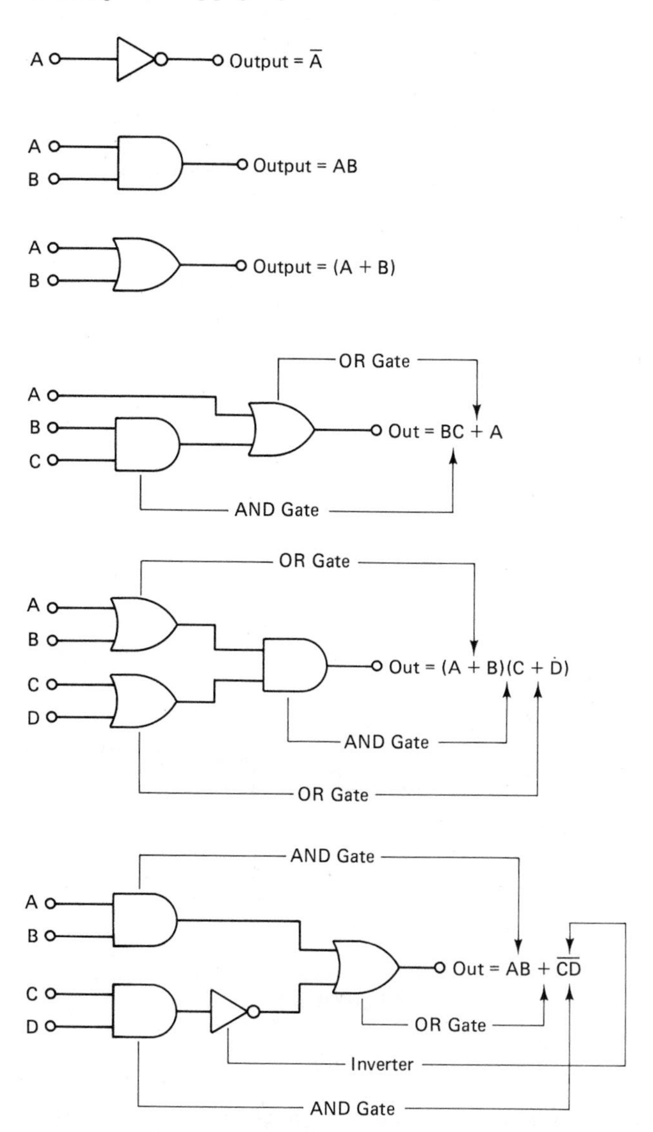
A
Output = $\overline{A}$
A
B
Output = AB
A
B
Output = (A + B)
OR Gate
A
B
C
Out = BC + A
AND Gate
OR Gate
A
B
C
D
Out = (A + B)(C + D)
AND Gate
OR Gate
AND Gate
A
B
C
D
Out = AB + $\overline{CD}$
OR Gate
Inverter
AND Gate

Figure 2-7

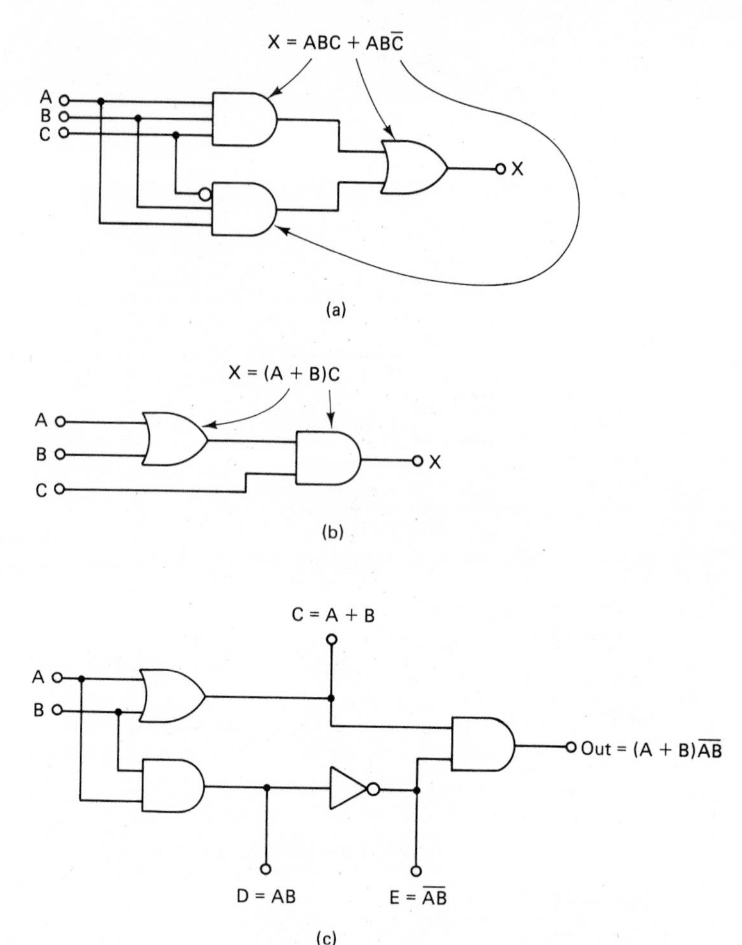

Figure 2-8

It is also possible to generate switching functions and logic circuits from a truth table. Since the truth table is the usual starting point when designing circuits, these techniques are very useful. As an example, consider the following:

Example 2-1:

Input → Two switches (A and B) that can be on (1) or off (0).
Output → A light that is either on (1) or off (0).

1. Design a logic circuit that will turn the light on when either switch is on but will shut the light off if both switches are on.
2. First construct a truth table consistent with all of the conditions. Since there are two inputs, there are $2^2 = 4$ combinations of inputs.

S_1	S_2
0	0
0	1
1	0
1	1

3. To meet the conditions, the light must be on with input conditions 01, 10. The complete truth table looks like this:

S_1	S_2	LIGHT
0	0	0
0	1	1
1	0	1
1	1	0

4. The light must be on when $S_1 = 0$, $S_2 = 1$ $(\overline{S}_1\ S_2)$ or $S_1 = 1$, $S_2 = 0$ $(S_1\ \overline{S}_2)$. The switching function must be:

$$\text{Light} = \overline{S}_1\ S_2 + S_1\ \overline{S}_2$$

5. The logic circuit looks like this:

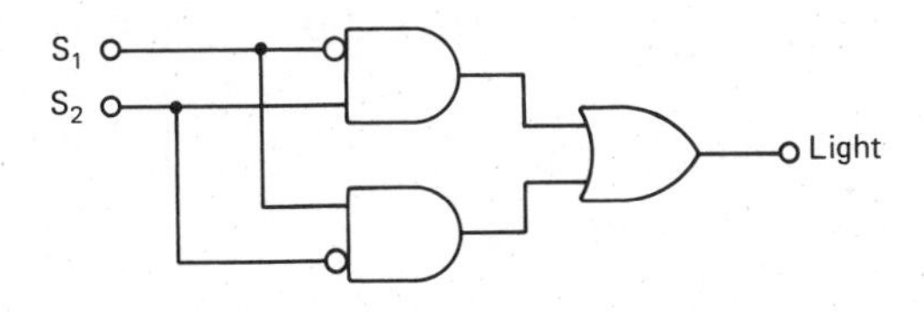

This method of deriving a switching function is known as sum-of-products. Indeed, a look at the switching function shows two products that are "summed."

SUM-OF-PRODUCTS

Given a truth table, a switching function can always be derived by using sum-of-products. The method obeys the following rules:

1. From the output column, pick out all of the "ones."
2. For each row that has a "one" as an output, write the product of the inputs using negation whenever an input is zero.
3. Take all of these products and add them together.

Example 2-2:

Given the following truth table, find the sum-of-products switching function (Figure 2-9):

A	B	C	OUT	
0	0	0	0	
0	0	1	1	*
0	1	0	0	
0	1	1	1	*
1	0	0	1	*
1	0	1	0	
1	1	0	1	*
1	1	1	0	

Figure 2-9

The "one" outputs are marked with asterisks. The products of these rows are:

$$\overline{A}\ \overline{B}\ C\ ,\ \overline{A}\ B\ C\ ,\ A\ \overline{B}\ \overline{C}\ ,\ A\ B\ \overline{C}$$

The switching function is:

$$OUT = \overline{A}\ \overline{B}\ C + \overline{A}\ B\ C + A\ \overline{B}\ \overline{C} + A\ B\ \overline{C}$$

A logic circuit for this function is shown in Figure 2-10.

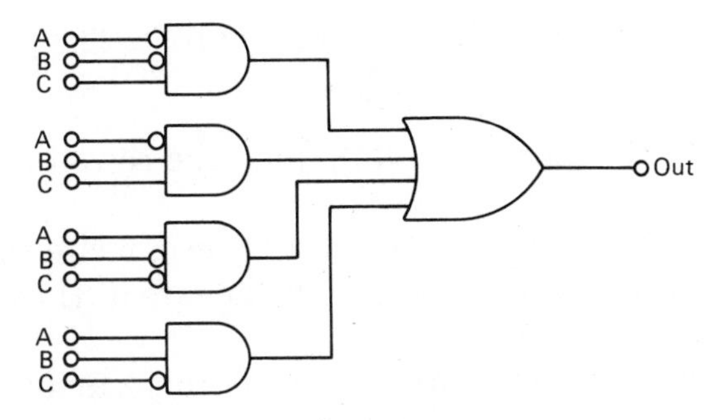

Figure 2-10

Each product represents an AND gate. The sum is represented by the OR gate on the output.

Example 2-3:

Find the sum-of-products switching function for the truth table seen in Figure 2-11.

A	B	OUT	
0	0	1	*
0	1	1	*
1	0	1	*
1	1	0	

Figure 2-11

The products are:

$\overline{A}\,\overline{B}$, $\overline{A}\,B$, $A\,\overline{B}$

Their sum → out $= \overline{A}\,\overline{B} + \overline{A}\,B + A\,\overline{B}$

PRODUCT-OF-SUMS

The product-of-sums method is in most ways the exact opposite of sum-of-products. Their results are identical since they both can be used on the same truth table. Algebraic manipulation of the functions to show that they give the same results are shown later in this chapter. The product-of-sums obeys the following rules:

1. From the output column, pick the rows that contain zero.
2. For each row whose output is zero, write the sum of the inputs. If the input is one negate it. If the input is zero write it non-negated.
3. Take all the sums and multiply them together.

Example 2-3:

Given the truth table in Figure 2-12, find the product-of-sums switching function.

A	B	C	OUT	
0	0	0	0	*
0	0	1	1	
0	1	0	0	*
0	1	1	1	
1	0	0	1	
1	0	1	0	*
1	1	0	1	
1	1	1	0	*

Figure 2-12

The zero outputs are marked with asterisks. The sums of these rows are:

$$(A + B + C), (A + \overline{B} + C), (\overline{A} + B + \overline{C}), (\overline{A} + \overline{B} + \overline{C})$$

The product-of-sums switching function is:

$$Out = (A + B + C)(A + \overline{B} + C)(\overline{A} + B + \overline{C})(\overline{A} + \overline{B} + \overline{C})$$

A logic circuit for this function is shown in Figure 2-13. Each sum represents an OR gate. The multiplication is represented by the AND gate on the output.

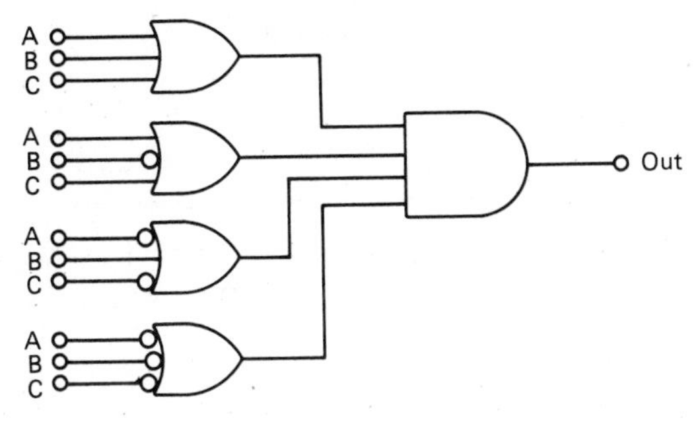

Figure 2-13

Both methods of deriving the switching function give equally valid results and it can be shown that the two expressions are equal. Which method is shorter depends on how many ones there are in the output column. Consider the truth table in Figure 2-14.

In this case, the sum-of-products is much easier and shorter since there are only two ones in the output.

A	B	C	OUT
0	0	0	0
0	0	1	0
0	1	0	1
0	1	1	0
1	0	0	0
1	0	1	1
1	1	0	0
1	1	1	0

Product-of-sums

$$\text{Out} = (A + B + C) \times (A + B + \overline{C})(A + \overline{B} + \overline{C}) \times (\overline{A} + B + C)(\overline{A} + \overline{B} + C) \times (\overline{A} + \overline{B} + \overline{C})$$

Sum-of-products

$$\text{Out} = \overline{A}B\overline{C} + A\overline{B}C$$

Figure 2-14

CHECKLIST OF BOOLEAN ALGEBRA RULES

The rules of Boolean Algebra are listed in Figure 2-15. They are useful in the manipulation and simplification of switching functions. Each rule can be understood by applying it to a logic function.

Boolean Algebra Rules

1.	$A \cdot 0 = 0$	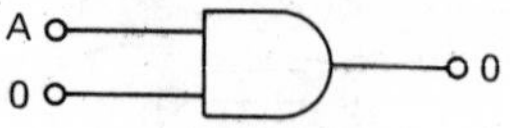
2.	$A \cdot 1 = A$	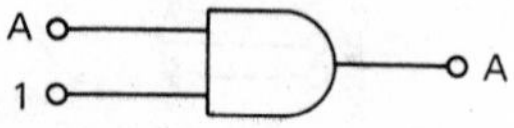
3.	$A \cdot A = A$	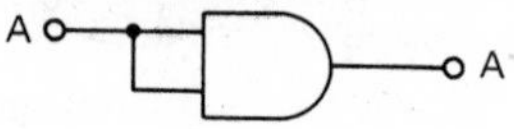
4.	$A \cdot \overline{A} = 0$	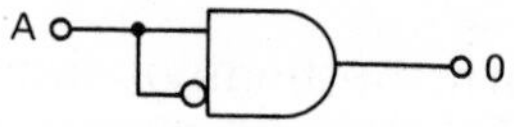
5.	$\overline{\overline{A}} = A$	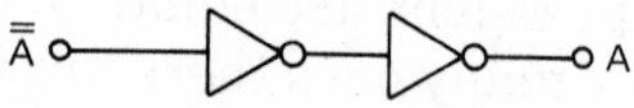
6.	$A + 0 = A$	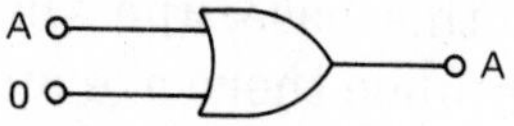
7.	$A + 1 = 1$	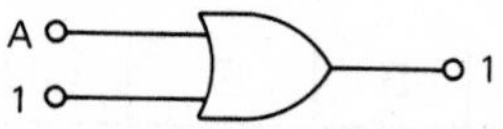
8.	$A + A = A$	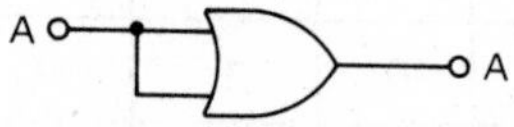
9.	$A + \overline{A} = 1$	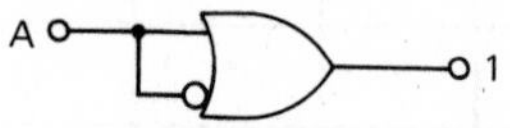
10.	$A + B = B + A$	Commutative Law
11.	$AB = BA$	Commutative Law
12.	$A + (B + C) = (A + B) + C$	Associative law

Figure 2-15

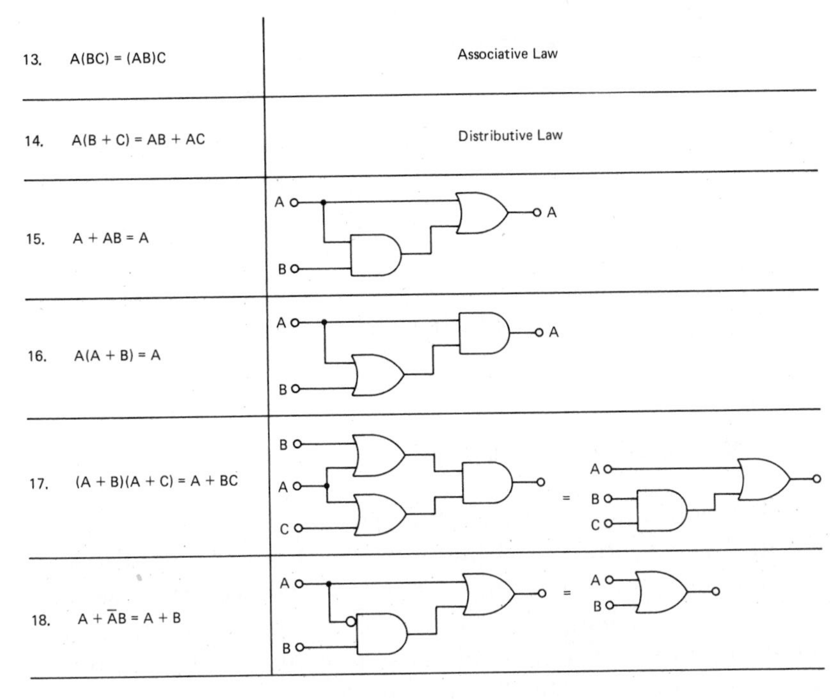

Figure 2-15 (continued)

Example:

Rule 3 says: A · A = A. This is an AND gate with A on both inputs.

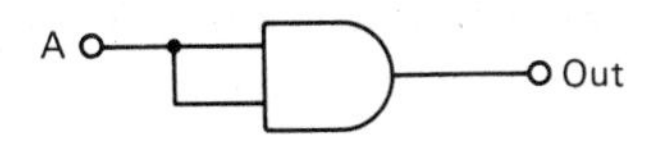

From the rules of an AND gate, if A = 1 the output is 1; if A = 0 the output is 0.

A · A = A

DEMORGAN'S THEOREM

The inversion can be changed from the output to the input or from input to output if the gate function is changed (Figure 2-16).

A B Equals A B

$\overline{A+B} = \bar{A}\bar{B}$

A B Equals A B

$\overline{AB} = \bar{A}+\bar{B}$

Figure 2-16

POSITIVE Vs NEGATIVE LOGIC

It is common in digital logic to use plus 5 volts as a binary one and zero volts as a binary zero. This is an example of positive logic since the "more positive" voltage represents a binary one. Although positive logic is more common, the choice of which voltage level to use to represent binary one is arbitrary. Letting the "more negative" voltage represent binary one is equally valid and this condition is called negative logic.

When the definition is changed from positive logic to negative logic, the basic gate functions are changed. AND gates become OR gates and OR gates become AND gates.

Example:

Consider the positive logic AND gate.

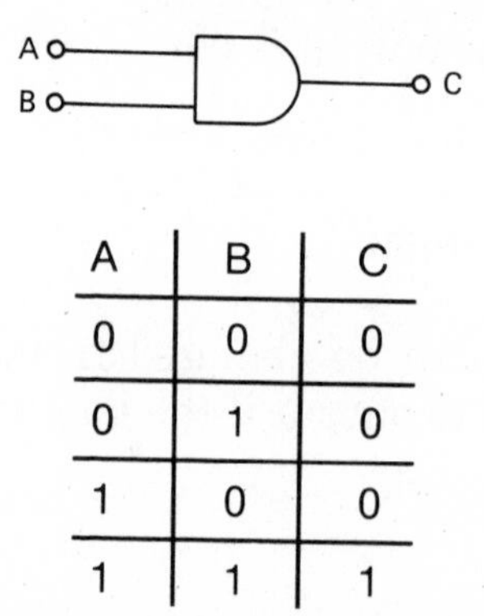

A	B	C
0	0	0
0	1	0
1	0	0
1	1	1

If the truth table is written in terms of voltage it becomes:

A	B	C
0V	0V	0V
0V	+5V	0V
+5V	0V	0V
+5V	+5V	+5V

If negative logic is now assumed, the zero volts become ones and the +5 volts become zeros. The truth table now becomes:

A	B	C
1	1	1
1	0	1
0	1	1
0	0	0

=

A
B
C

This is the truth table for an OR gate. If the same procedure is used beginning with an OR gate, the function will change to AND.

IMPLIED GATES

Some logic diagrams contain symbols like those in Figure 2-17. They represent an implied AND or OR gating function. The switching function is taking place simply by connecting the outputs of certain gates together. The gates are not physically in the circuit in the form of components but are implied. The output result is the same.

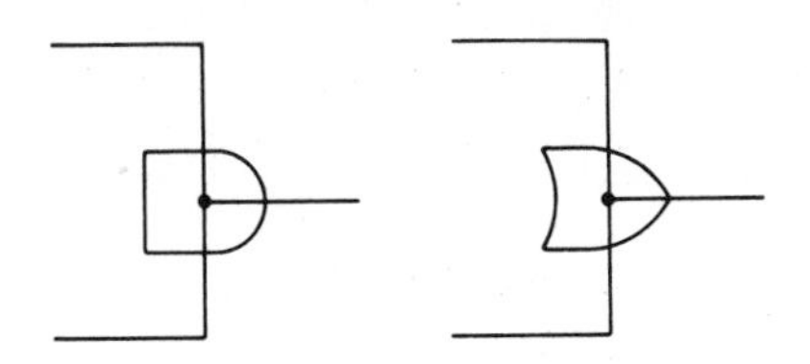

Figure 2-17

HOW TO SIMPLIFY AND MANIPULATE EQUATIONS

Using the rules of Boolean Algebra and DeMorgan's theorem, switching functions can be simplified and manipulated. It is possible to minimize the number of gates used, or to rearrange the gates to use the type of gate desired. The following examples show how this is done.

Example 1:

Change the form of the switching function $C = (A + \overline{B}) \times (\overline{A} + B)$ (Figure 2-18).

Multiply the two sums (Figure 2-19).

$$\begin{array}{r} A + \overline{B} \\ \times\ \overline{A} + B \\ \hline \end{array}$$

$\overline{A}A + \overline{A}\overline{B} + AB + B\overline{B}$ Multiply

$\overline{B}B = \overline{A}A = 0$ Rule 4

$C = \overline{A}\overline{B} + AB$

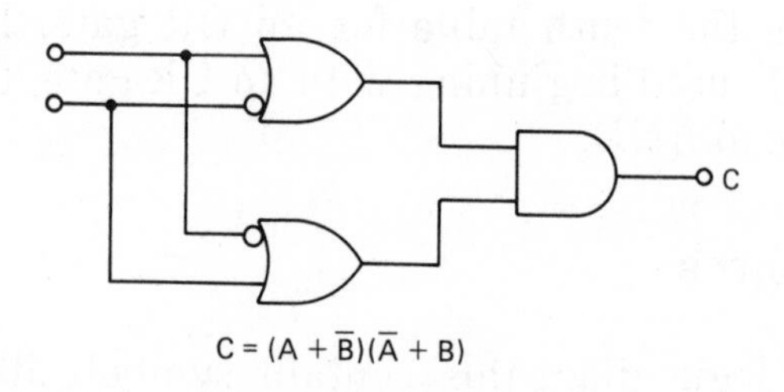

$C = (A + \overline{B})(\overline{A} + B)$

Figure 2-18

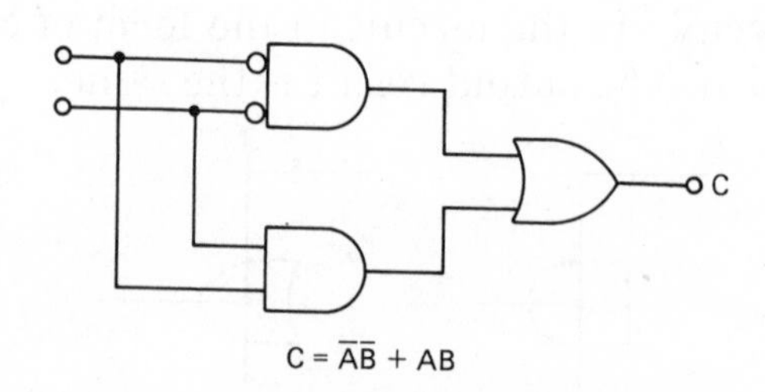

$C = \overline{A}\overline{B} + AB$

Figure 2-19

This is an example of rearranging the switching function. The rearranged circuit is not simpler, it is another circuit to implement the same function.

Example 2:

Implement the switching function X = AB + BC using NAND gates (Figure 2-20).

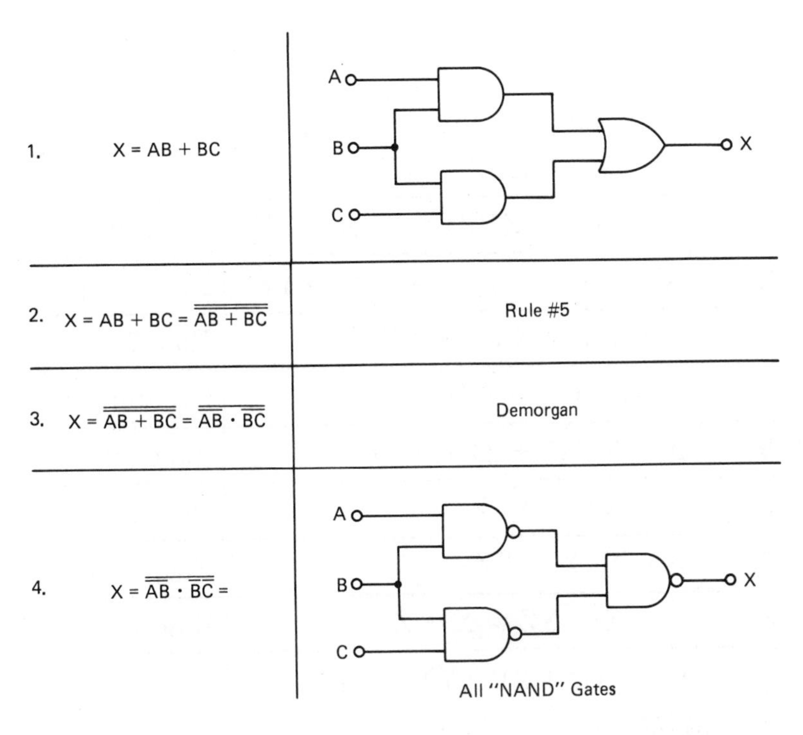

Figure 2-20

To simplify switching functions by using algebraic methods, it is necessary to observe the following rules:

1. Wherever possible, perform the indicated multiplication.
2. Look for product terms where one, and only one, term changes, such as:

$$\overline{X}\,Y + X\,Y$$

$$Y\,(\overline{X} + X)$$

$$Y$$

3. Look for product terms that can be factored such as:

$$X\,Y + X\,Y\,Z$$

$$X\,Y\,(1 + Z)$$

$$XY$$

Example 3:

Simplify the following switching function.

	$A = XY + XYZ + XY\overline{Z} + \overline{X}YZ$	
1	$A = (XY + XYZ) + XY\overline{Z} + \overline{X}YZ$	Grouping
2	$A = XY\,(1 + Z) + XY\overline{Z} + \overline{X}YZ$	Factoring
3	$A = XY + XY\overline{Z} + \overline{X}YZ$	Rule 7
4	$A = (XY + XY\overline{Z}) + \overline{X}YZ$	Grouping
5	$A = XY\,(1 + \overline{Z}) + \overline{X}YZ$	Factoring
6	$A = XY + \overline{X}YZ$	Rule 7
7	$A = Y\,(X + \overline{X}Z)$	Factoring
8	$X + \overline{X}Z = X + Z$	Rule 18
9	$A = Y\,(X + Z) = YX + YZ$	

Example 4:

For the following truth table, derive the simplest switching function.

A	B	OUT
0	0	1
0	1	1
1	0	0
1	1	0

USING PRODUCT-OF-SUMS

$$OUT = (\overline{A} + B)(\overline{A} + \overline{B})$$
$$(\overline{A} + B)(\overline{A} + \overline{B}) = \overline{A}\,\overline{A} + \overline{A}B + \overline{A}\,\overline{B} + B\overline{B}$$
$$\overline{A}\,\overline{A} = \overline{A}$$
$$B\overline{B} = 0$$
$$OUT = \overline{A} + \overline{A}B + \overline{A}\,\overline{B}$$
$$OUT = \overline{A}\,(1 + B + \overline{B})$$
$$OUT = \overline{A}$$

USING SUM-OF PRODUCTS

$$OUT = \overline{A}\,\overline{B} + \overline{A}B$$
$$OUT = \overline{A}\,(\overline{B} + B)$$
$$OUT = \overline{A}$$

USING MAPS TO SIMPLIFY

Switching functions can be simplified and manipulated using Boolean Algebra rules and DeMorgan's theorem. If the goal is to simplify the function, it is more common to use one of the graphic techniques. The techniques of this section, n-cube, Karnaugh maps, and Quine-McCluskey, have different advantages in the simplification of switching functions.

N-CUBE

Given a binary variable B, this variable can only have two values. These values are zero and one. This can be represented by a line segment with the two ends of the line representing the values of zero and one (Figure 2-21).

B = 0 ●————————● B = 1

Figure 2-21

If one variable can be represented by a line segment, two variables can be represented by a square (Figure 2-22).

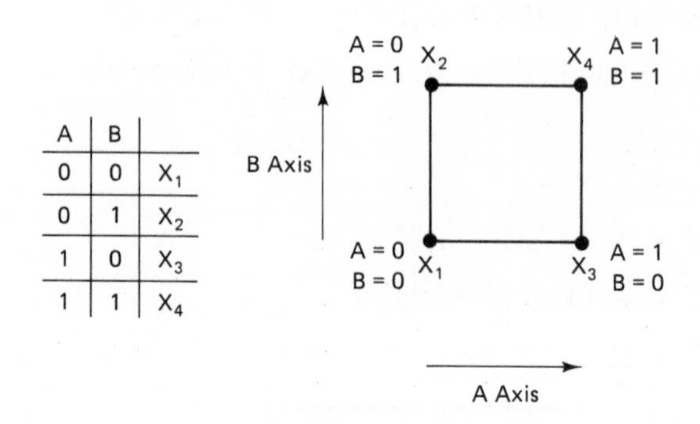

A	B	
0	0	X_1
0	1	X_2
1	0	X_3
1	1	X_4

Figure 2-22

With two variables, there are four possible combinations of these variables as shown in Figure 2-22. The four vertices of a square can be used to represent these four combinations. On the square the horizontal axis represents the variable A. The vertical axis represents the variable B.

With three variables, there are eight possible combinations. These combinations can be represented by the vertices of a cube as shown in Figure 2-23. The A, B, and C variables are shown on the eight vertices. The vertical axis represents the B variable, the horizontal axis the A variable, and the depth axis represents the C variable.

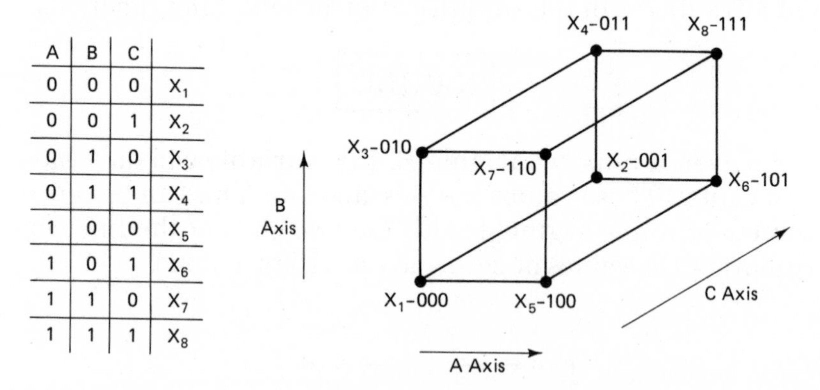

A	B	C	
0	0	0	X_1
0	0	1	X_2
0	1	0	X_3
0	1	1	X_4
1	0	0	X_5
1	0	1	X_6
1	1	0	X_7
1	1	1	X_8

Figure 2-23

The same concept can be used to extend the n-cube representation to four variables. The four-variable cube, however, is cumbersome. It involves a four-dimensional hypercube as shown in Figure 2-24. The configuration is a cube inside a larger cube with the vertices of the two cubes connected. The horizontal axis represents the variable A. The vertical axis represents B. The depth axis represents C, and the lines connecting the vertices between cubes represents D. The hypercube has a total of 16 vertices.

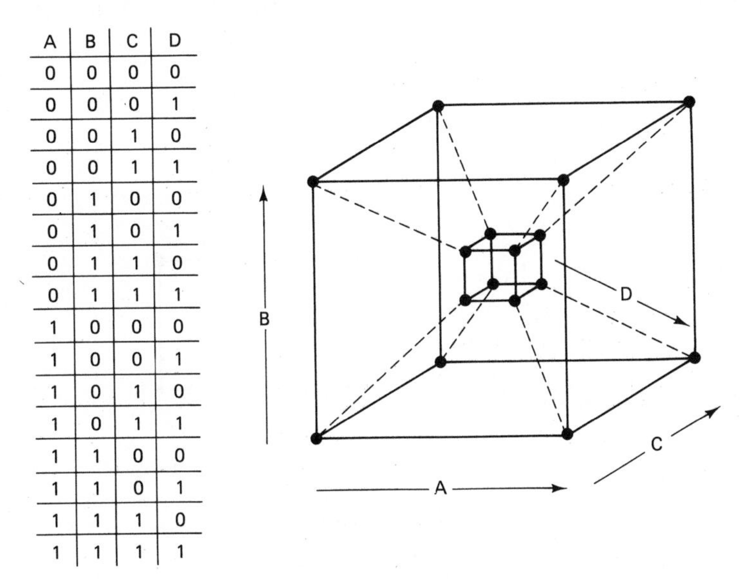

A	B	C	D
0	0	0	0
0	0	0	1
0	0	1	0
0	0	1	1
0	1	0	0
0	1	0	1
0	1	1	0
0	1	1	1
1	0	0	0
1	0	0	1
1	0	1	0
1	0	1	1
1	1	0	0
1	1	0	1
1	1	1	0
1	1	1	1

Figure 2-24

HOW TO SIMPLIFY

The n-cube provides a quick visual clue as to whether a switching function can be simplified. It does so by the following rules:

Rule 1:

If adjacent vertices on the n-cube have dots, the function can be simplified. If there are no adjacent dots, the function cannot be simplified.

Rule 2:

If adjacent vertices have dots, the variable that changed values along that line can be eliminated from the switching function.

These rules can best be illustrated by an example. Consider the truth table in Figure 2-25. The n-cube has two adjacent vertices with dots, therefore, the function can be simplified. The variable that changed along the line connecting the DOTS is B. Therefore, B can be eliminated from the switching function. The switching function then becomes:

$$\boxed{C = A}$$

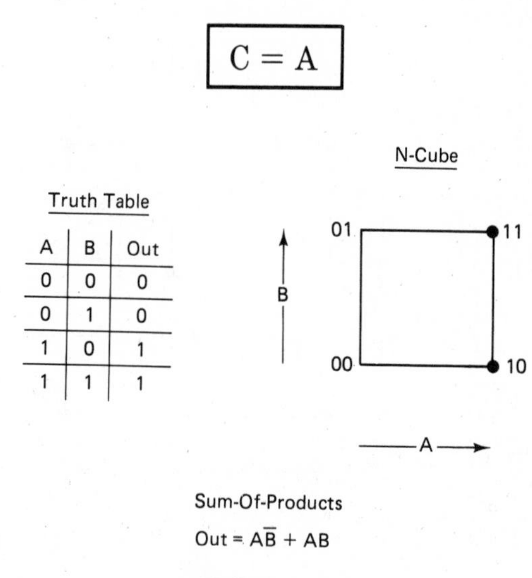

Figure 2-25

Example 2:

Consider the truth table, N-cube, and switching function in Figures 2-26.

Each product term in the switching function corresponds to a dot on the vertices of the N-cube (Figure 2-27).

This N-cube has two adjacent vertices with dots. By the rules of simplification, the two terms can be

A	B	C	Out
0	0	0	0
0	0	1	1
0	1	0	0
0	1	1	0
1	0	0	1
1	0	1	0
1	1	0	1
1	1	1	0

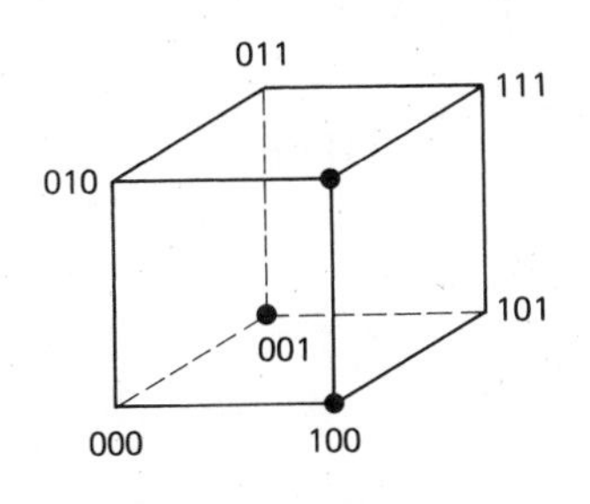

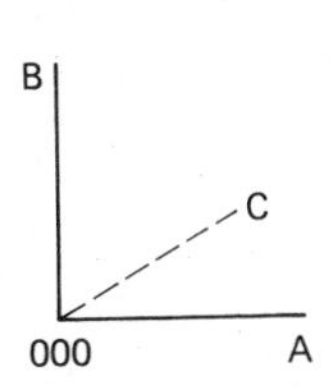

Figure 2-26

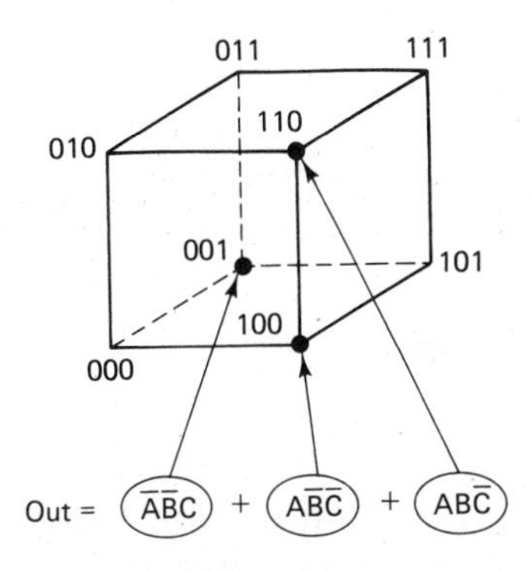

Figure 2-27

combined and the variable that changed between the dots can be eliminated.

$$\text{OUT} = \overline{A}\,\overline{B}\,C + A\,\overline{C}$$

combination of two terms with B eliminated.

Example 3:

It is possible to have more than two adjacent vertices with dots. In this case, the only remaining term is the variable that did not change (Figure 2-28).

This example has four adjacent vertices with dots. By the rules of simplification, all of the variables that changed can be eliminated. The only variable that did not change is B. The simplified switching function is then:

$$OUT = B$$

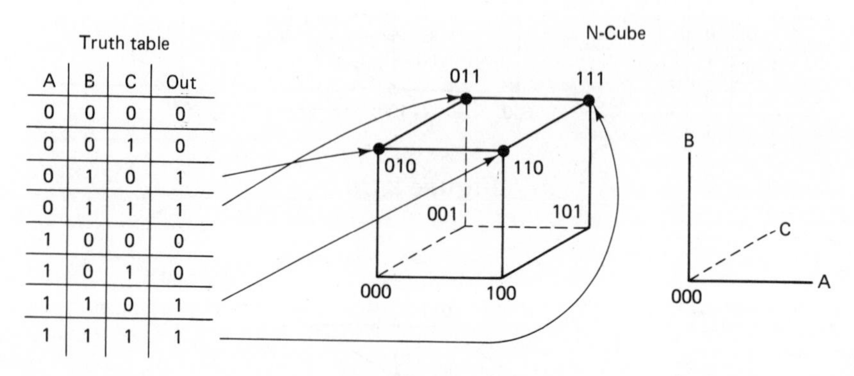

Figure 2-28

THE KARNAUGH MAP

The Karnaugh map is another graphic method by which a switching function can be simplified. Whereas the N-cube represented each possible truth table output as the vertices of a cube, the Karnaugh map represents truth table outputs as small sections of a rectangle. The layout of the Karnaugh map is such that it allows simplification of the switching function. Karnaugh maps for two-, three-, four-, and five-variable maps are shown in Figures (2-29) through (2-32).

Truth Table

A	B	OUT
0	0	X_1
0	1	X_2
1	0	X_3
1	1	X_4

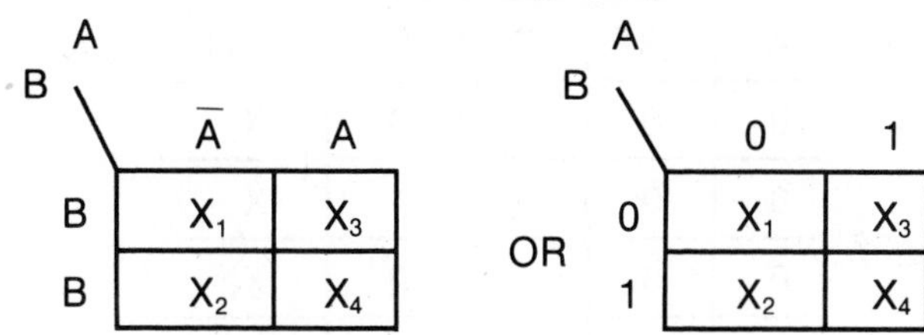

Figure 2-29

Truth Table

A	B	C	OUT
0	0	0	X_1
0	0	1	X_2
0	1	0	X_3
0	1	1	X_4
1	0	0	X_5
1	0	1	X_6
1	1	0	X_7
1	1	1	X_8

Karnaugh Map

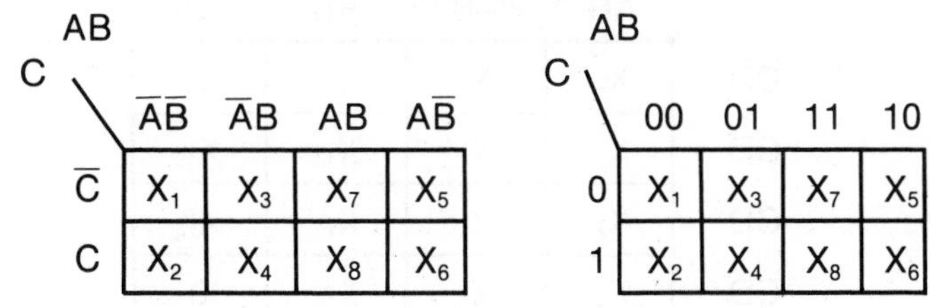

Figure 2-30

Truth Table

A	B	C	D	OUT
0	0	0	0	X_1
0	0	0	1	X_2
0	0	1	0	X_3
0	0	1	1	X_4
0	1	0	0	X_5
0	1	0	1	X_6
0	1	1	0	X_7
0	1	1	1	X_8
1	0	0	0	X_9
1	0	0	1	X_{10}
1	0	1	0	X_{11}
1	0	1	1	X_{12}
1	1	0	0	X_{13}
1	1	0	1	X_{14}
1	1	1	0	X_{15}
1	1	1	1	X_{16}

Figure 2-30 (continued)

Karnaugh Map

CD \ AB	$\overline{A}\overline{B}$	$\overline{A}B$	AB	$A\overline{B}$
$\overline{C}\overline{D}$	X_1	X_5	X_{13}	X_9
$\overline{C}D$	X_2	X_6	X_{14}	X_{10}
CD	X_4	X_8	X_{16}	X_{12}
$C\overline{D}$	X_3	X_7	X_{15}	X_{11}

Figure 2-31

CD \ AB	00	01	11	10
00	X_1	X_5	X_{13}	X_9
01	X_2	X_6	X_{14}	X_{10}
11	X_4	X_8	X_{16}	X_{12}
10	X_3	X_7	X_{15}	X_{11}

Figure 2-31 (continued)

Truth Table

A	B	C	D	E	OUT
0	0	0	0	0	X_1
0	0	0	0	1	X_2
0	0	0	1	0	X_3
0	0	0	1	1	X_4
0	0	1	0	0	X_5
0	0	1	0	1	X_6
0	0	1	1	0	X_7
0	0	1	1	1	X_8
0	1	0	0	0	X_9
0	1	0	0	1	X_{10}
0	1	0	1	0	X_{11}
0	1	0	1	1	X_{12}
0	1	1	0	0	X_{13}
0	1	1	0	1	X_{14}
0	1	1	1	0	X_{15}

Figure 2-32

0	1	1	1	1	X_{16}
1	0	0	0	0	X_{17}
1	0	0	0	1	X_{18}
1	0	0	1	0	X_{19}
1	0	0	1	1	X_{20}
1	0	1	0	0	X_{21}
1	0	1	0	1	X_{22}
1	0	1	1	0	X_{23}
1	0	1	1	1	X_{24}
1	1	0	0	0	X_{25}
1	1	0	0	1	X_{26}
1	1	0	1	0	X_{27}
1	1	0	1	1	X_{28}
1	1	1	0	0	X_{29}
1	1	1	0	1	X_{30}
1	1	1	1	0	X_{31}
1	1	1	1	1	X_{32}

Karnaugh Map

DE \ ABC	000	001	011	010	110	111	101	100
00	X_1	X_5	X_{13}	X_9	X_{25}	X_{29}	X_{21}	X_{17}
01	X_2	X_6	X_{14}	X_{10}	X_{26}	X_{30}	X_{22}	X_{18}
11	X_4	X_8	X_{16}	X_{12}	X_{28}	X_{32}	X_{24}	X_{20}
10	X_3	X_7	X_{15}	X_{11}	X_{27}	X_{31}	X_{23}	X_{19}

Figure 2-32 (continued)

The Two-Variable Map

The two-variable Karnaugh map has one input on the vertical axis, one input on the horizontal axis, and the four outputs in the segments of the rectangle.

Three-Variable Map

The three-variable map has one input on the vertical axis and two inputs on the horizontal axis as shown in Figure 2-30.

Four-Variable Map

The four-variable map has two inputs on the vertical axis and two inputs on the horizontal axis as shown in Figure 2-31.

Five-Variable Map

The five-variable Karnaugh map can be constructed in many different ways. Most of the five-variable maps become very cumbersome and hard to use for simplification. One form of the five-variable map is shown in Figure 2-32.

Note 1: The vertical and horizontal scales of the Karnaugh map are written in "Gray" code. One, and only one, variable changes in subsequent steps. Ones in adjacent cells of the map indicate that the switching function can be simplified.

Note 2: Karnaugh maps are considered to be "rolled." The top and bottom lines are adjacent and the left-hand and right-hand sides are adjacent. See Figure 2-33.

SIMPLIFYING

Rule: If ones appear in adjacent cells of the Karnaugh map, the switching function can be simplified (Figure 2-34).

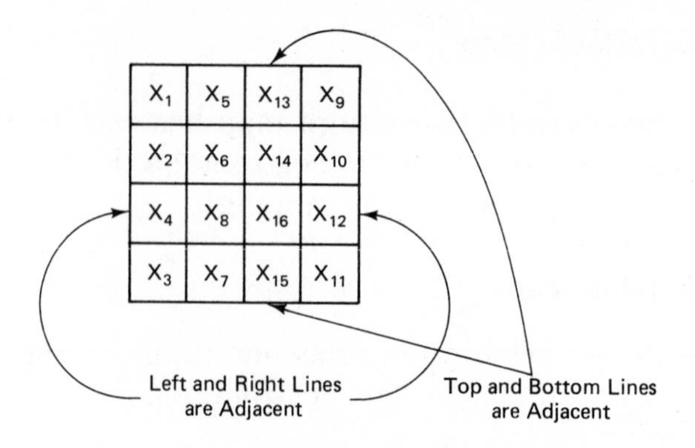

Figure 2-33

Truth Table

A	B	OUT
0	0	1
0	1	1
1	0	0
1	1	0

Karnaugh Map

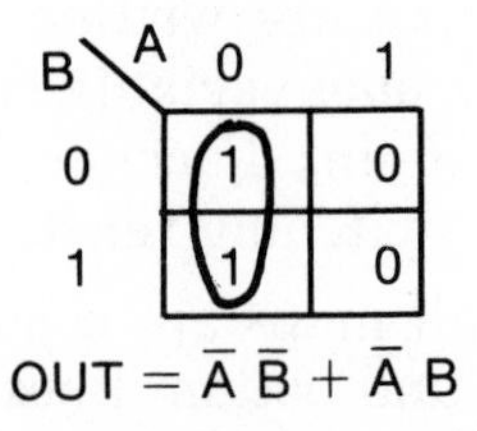

Figure 2-34

Example 1:

The Karnaugh map has ones in adjacent cells (circled area in the map, Figure 2-34; therefore, the switching function can be simplified. The variable that changed between the adjacent cells is B; therefore, B can be eliminated. The simplified switching function becomes:

$$OUT = \overline{A}$$

The same results can be obtained by algebraic simplification.

$$OUT = \overline{A}\,\overline{B} + \overline{A}\,B$$
$$OUT = \overline{A}\,(\overline{B} + B) \longrightarrow \text{Factoring}$$
$$\overline{B} + B = 1 \longrightarrow \text{Rule 9}$$
$$OUT = \overline{A}$$

Example 2:

A more complex example is given in Figure 2-35.

The map has three adjacent ones. These three terms can be combined leaving only the variable that did not change. The simplified switching function is:

$$OUT = \overline{A}\,\overline{B}\,C + \overline{C}$$

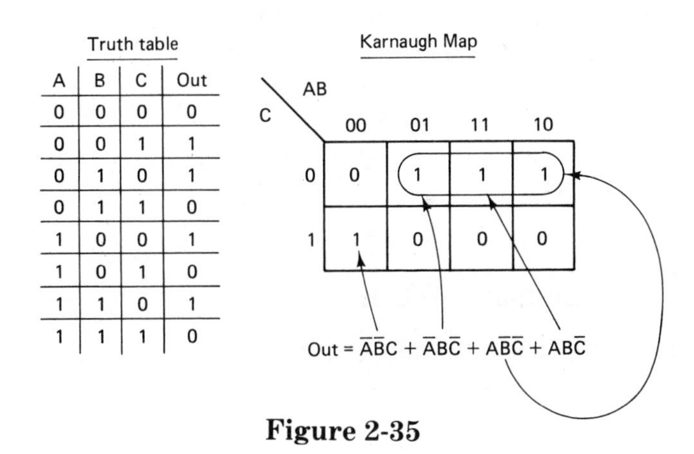
Truth table

A	B	C	Out
0	0	0	0
0	0	1	1
0	1	0	1
0	1	1	0
1	0	0	1
1	0	1	0
1	1	0	1
1	1	1	0

Figure 2-35

Example 3:

It is also possible to have groupings of adjacent ones intersect (Figure 2-36).

The simplified switching function is written as one term for each circle in the Karnaugh map, while eliminating the variable that changed.

$$OUT = B + A \longleftarrow \text{Circle to the Right}$$

↖Lower Circle (under B)

Truth table

A	B	Out
0	0	0
0	1	1
1	0	1
1	1	1

Karnaugh Map

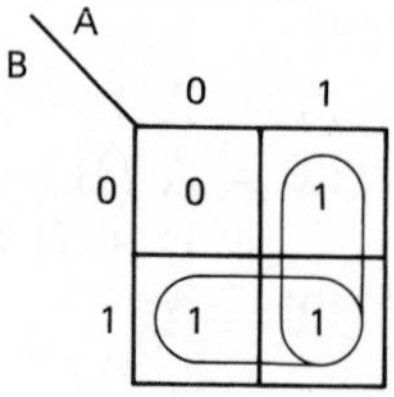

B \ A	0	1
0	0	1
1	1	1

$Out = \overline{A}B + A\overline{B} + AB$

Figure 2-36

Example 4:

The map in Figure 2-37 has adjacent ones because of its rolled condition and also has a common cell. Taking the rolled condition first, the variable that changed was A. The two terms can be combined to:

$$\overline{B}\ C\ \overline{D}$$

Taking the intersecting circles, a term can be written for each circle.

$$\text{Left} = \overline{A}\ B\ D$$

$$\text{Bottom} = B\ C\ D$$

The simplified switching function is:

$$OUT = \overline{B}\ C\ \overline{D} + \overline{A}\ B\ D + B\ C\ D$$

THE QUINE McCLUSKEY METHOD

The Quine McCluskey method (also called tabulation) is an important simplification method for two reasons.

1. It can be used with any number of variables. Most map methods become cumbersome if more than four variables are used.

2. It is a systematic method that can be programmed on a computer.

Truth table

A	B	C	D	Out
0	0	0	0	0
0	0	0	1	0
0	0	1	0	1
0	0	1	1	0
0	1	0	0	0
0	1	0	1	1
0	1	1	0	0
0	1	1	1	1
1	0	0	0	0
1	0	0	1	0
1	0	1	0	1
1	0	1	1	0
1	1	0	0	0
1	1	0	1	0
1	1	1	0	0
1	1	1	1	1

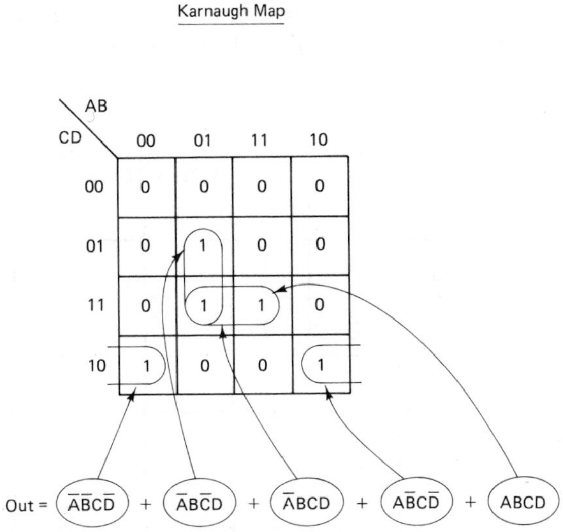

Karnaugh Map

CD \ AB	00	01	11	10
00	0	0	0	0
01	0	1	0	0
11	0	1	1	0
10	1	0	0	1

$$\text{Out} = \overline{A}\,\overline{B}C\overline{D} + \overline{A}B\overline{C}D + \overline{A}BCD + A\overline{B}C\overline{D} + ABCD$$

Figure 2-37

Consider the truth table in Figure 2-38.
The sum-of-products switching function is:

$$\begin{aligned}\text{OUT} = {} & \overline{A}\,\overline{B}\,\overline{C}\,\overline{D} + \overline{A}\,\overline{B}C\overline{D} + \\ & \overline{A}BC\overline{D} + \overline{A}BCD + \\ & A\overline{B}C\overline{D} + AB\overline{C}D + \\ & ABC\overline{D} + ABCD\end{aligned}$$

If we label each row of the truth table with its corresponding decimal number (0 through 15), we can also write the switching function using the following notation:

$$\text{OUT} = \Sigma\ 0,2,6,7,10,13,14,15$$

$$\begin{aligned}\text{WHERE } 0 &= \overline{A}\,\overline{B}\,\overline{C}\,\overline{D} \\ 2 &= \overline{A}\,\overline{B}C\overline{D} \\ &\text{etc.}\end{aligned}$$

This type of notation is a common way to express a switching function when using the Quine McCluskey method.

Truth Table

A	B	C	D	OUT	DECIMAL
0	0	0	0	1	0
0	0	0	1	0	1
0	0	1	0	1	2
0	0	1	1	0	3
0	1	0	0	0	4
0	1	0	1	0	5
0	1	1	0	1	6
0	1	1	1	1	7
1	0	0	0	0	8
1	0	0	1	0	9
1	0	1	0	1	10
1	0	1	1	0	11
1	1	0	0	0	12
1	1	0	1	1	13
1	1	1	0	1	14
1	1	1	1	1	15

Figure 2-38

To simplify the switching function using Quine McCluskey, you first arrange the products in order of the number of "ones" as shown in Figure 2-39.

$$\Sigma\ 0, 2, 6, 7, 10, 13, 14, 15$$

	Σ		# OF ONES
$\overline{A}\,\overline{B}\,C\,\overline{D}$	0	0000	0
$\overline{A}\,\overline{B}\,C\,\overline{D}$	2	0010	1
$\overline{A}\,B\,C\,\overline{D}$	6	0110	2
$A\,\overline{B}\,C\,\overline{D}$	10	1010	
$\overline{A}\,B\,C\,D$	7	0111	3
$A\,B\,\overline{C}\,D$	13	1101	
$A\,B\,C\,\overline{D}$	14	1110	
$A\,B\,C\,D$	15	1111	4

Figure 2-39

The table in Figure 2-39 is reduced according to the following rules.

1. Starting with the row containing the least number of ones, compare it with each row in the next group below. Each comparison showing one, and only one, bit changing is reduced according to Figure 2-40. The bit that changed is denoted by a dash mark (-).

A	B	C	D	
0	0	0	0	√
0	0	1	0	√
0	1	1	0	√
1	0	1	0	√
0	1	1	1	√
1	1	0	1	√
1	1	1	0	√
1	1	1	1	√

First Reduction

A	B	C	D
0	0	—	0
0	—	1	0
—	0	1	0
0	1	1	—
—	1	1	0
1	—	1	0
—	1	1	1
1	1	—	1
1	1	1	—

Figure 2-40

2. Each time a pair is reduced, a check mark (√) is placed beside each number as shown in Figure 2-40.

The first reduction shown in Figure (2-40) is also arranged in groups according to the number of ones. In this example, each term in the first table was able to combine and there is a check (√) beside each one. This is not always the case. This reduction technique is continued until no further reduction can take place. When going from the first reduction to the second reduction, the (—) marks must match before the two terms can be combined (Figure 2-41).

First Reduction

A	B	C	D	
0	0	—	0	
0	—	1	0	√
—	0	1	0	√
0	1	1	—	√
—	1	1	0	√
1	—	1	0	√
—	1	1	1	√
1	1	—	1	
1	1	1	—	√

Second Reduction

A	B	C	D
—	—	1	0
—	1	1	—

Figure 2-41

An examination of the second reduction shows that it can be reduced no further. An examination also shows that two rows in the first reduction and the two rows in the second reduction are not checked. This is because they did not combine with any other terms.

A reduced switching function can be made from the terms with no check marks.

$$\text{OUT} = \overline{A}\,\overline{B}\,\overline{D} + A\,B\,D + C\,\overline{D} + B\,C$$

This function is a simplified function for the truth table we started with. It may or may not be the simplest form. In order to determine if it is the simplest form, we must construct another table (Figure 2-42). This table compares the original switching function terms to the simplified switching function terms.

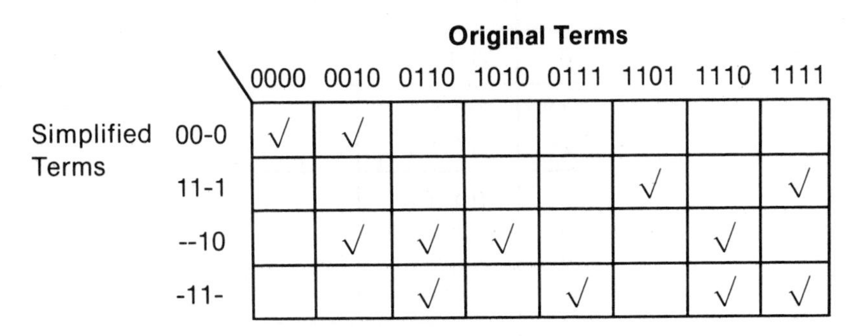

Simplified Terms \ Original Terms	0000	0010	0110	1010	0111	1101	1110	1111
00-0	√	√						
11-1						√		√
--10		√	√	√			√	
-11-			√		√		√	√

Figure 2-42

A check mark is placed in the square each time the simplified term "covers" an original term. The dash (-) marks can stand for either zero or one. For example --10 can cover 1110, 1010, 0110, 0010. An examination of the table shows, that if any of the simplified terms were eliminated, one or more of the original terms would not be covered. This means that the switching function cannot be simplified further. The simplest function is:

$$OUT = \overline{A}\ \overline{B}\ \overline{D} + A\ B\ D + C\ \overline{D} + B\ C$$

Example 1:

By Quine McCluskey, simplify the following switching function.

$$\Sigma\ 0, 1, 2, 5, 6, 7, 9, 10, 11, 13, 14, 15$$

First arrange the products according to the number of ones (Figure 2-43). See Figure 2-44 for first and second reductions.

The second reduction cannot be reduced any further. The terms left with no check marks form the following simplified switching function:

$$OUT = \overline{A}\,\overline{B}\,\overline{C} + \overline{A}\,\overline{B}\,\overline{D} + \overline{C}\,D + C\,\overline{D} + B\,D + B\,C + A\,D + A\,C$$

Σ	A B C D	# OF ONES
0	0 0 0 0	0
1	0 0 0 1	1
2	0 0 1 0	
5	0 1 0 1	2
6	0 1 1 0	
9	1 0 0 1	
10	1 0 1 0	
7	0 1 1 1	3
11	1 0 1 1	
13	1 1 0 1	
14	1 1 1 0	
15	1 1 1 1	4

Figure 2-43

A	B	C	D	
0	0	0	0	√
0	0	0	1	√
0	0	1	0	√
0	1	0	1	√
0	1	1	0	√
1	0	0	1	√
1	0	1	0	√
0	1	1	1	√
1	0	1	1	√
1	1	0	1	√
1	1	1	0	√
1	1	1	1	√

Figure 2-44

First Reduction

A	B	C	D	
0	0	0	—	
0	0	—	0	
0	—	0	1	√
—	0	0	1	√
0	—	1	0	√
—	0	1	0	√
0	1	—	1	√
—	1	0	1	√
0	1	1	—	√
—	1	1	0	√
1	0	—	1	√
1	—	0	1	√
1	0	1	—	√
1	—	1	0	√
—	1	1	1	√
1	—	1	1	√
1	1	—	1	√
1	1	1	—	√

Second Reduction

A	B	C	D	
—	—	0	1 }	★
—	—	0	1 }	
—	—	1	0 }	★
—	—	1	0 }	
—	1	—	1 }	★
—	1	—	1 }	
—	1	1	—	
1	—	—	1	
1	—	1	—	

Figure 2.44 (continued)

*When reduced terms turn out the same, only one needs to be considered since A + A = A.

To determine if this is the simplest function, it is necessary to chart the simplified terms against the original switching function terms. (See Figure 2-45.) The original terms were:

$$\Sigma\ 0, 1, 2, 5, 6, 7, 9, 10, 11, 13, 14, 15$$

$$\begin{aligned} OUT = {} & \overline{A}\,\overline{B}\,\overline{C}\,\overline{D} + \overline{A}\,\overline{B}\,\overline{C}\,D + \overline{A}\,\overline{B}\,C\,\overline{D} + \overline{A}\,B\,\overline{C}\,D + \\ & \overline{A}\,B\,C\,\overline{D} + \overline{A}\,B\,C\,D + A\,\overline{B}\,\overline{C}\,D + A\,\overline{B}\,C\,\overline{D} + \\ & A\,\overline{B}\,C\,D + A\,B\,\overline{C}\,D + A\,B\,C\,\overline{D} + A\,B\,C\,D \end{aligned}$$

An examination of this chart shows that the switching function has not been minimized. The simplified products on the vertical axis marked with an asterisk are all covered by the other simplified products. These terms can be eliminated and the simplified switching function becomes:

$$OUT = \overline{A}\,\overline{B}\,\overline{D} + \overline{C}\,D + B\,D + A\,C$$

EXAMPLES OF SOLVED PROBLEMS

The following examples represent the basic method of deriving a logic circuit that will perform a predetermined function.

Example 1:

Design a logic circuit to control a lamp. The lamp is to be turned on or off according to the following conditions.

1. The lamp will always be on between 6:00 p.m. and 6:00 a.m. regardless of other conditions.
2. The lamp will go off if lamp X is turned on except per number 1.

Let (A) represent the time → 1 for 6:00 p.m. to 6:00 a.m. and 0 for daytime.
Let (B) represent lamp X → 1 when turned on and 0 when turned off.

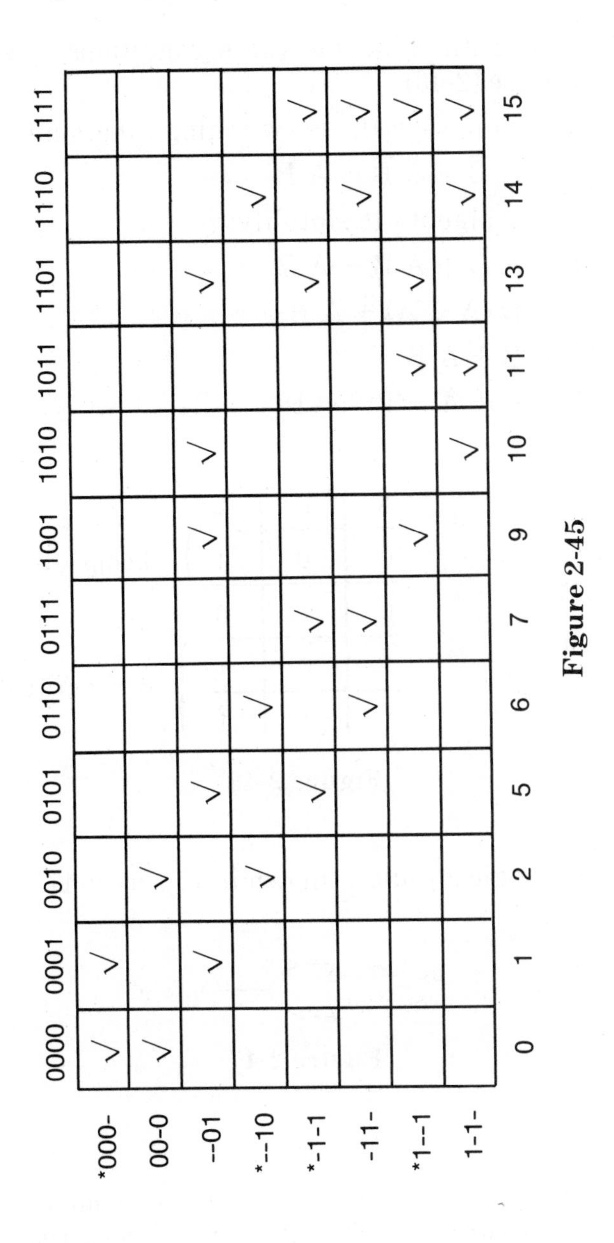

	0000	0001	0010	0101	0110	0111	1001	1010	1011	1101	1110	1111
*000-	√	√										
00-0	√		√									
--01		√		√			√	√		√		
*--10			√		√						√	
*-1-1				√		√				√		√
-11-					√	√					√	√
*1--1							√		√	√		√
1-1-								√	√		√	√
	0	1	2	5	6	7	9	10	11	13	14	15

Figure 2-45

A. The truth table for these conditions is shown in Figure (2-46)

B. The sum-of-products switching function is:

$$L = \overline{A}\,\overline{B} + A\,\overline{B} + A\,B$$

C. Using algebra to simplify:

$$L = \overline{A}\,\overline{B} + A\,\overline{B} + A\,B$$

$$L = \overline{B}\,(\overline{A} + A) + A\,B$$

$$L = \overline{B} + A\,B$$

$$L = B + A \rightarrow \text{Rule \#18}$$

A	B	L	
0	0	1	Lamp X Is Off
0	1	0	
1	0	1	6:00 pm To 6:00 am
1	1	1	

Figure 2-46

D. The logic circuit is shown in Figure 2-47.

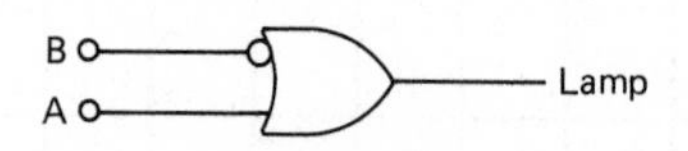

Figure 2-47

Example 2:

Design a logic circuit that will control a lamp. There are three switches. Any one of the three switches can turn the lamp on or off regardless of its original state.

A. The first step is to derive a truth table. Let 1 represent the lamp "on" and (0) represent the lamp "off". Since there are three inputs (three switches) there must be eight combinations of inputs (see Figure 2-48).

S_1	S_2	S_3	Lamp
0	0	0	
0	0	1	
0	1	0	
0	1	1	
1	0	0	
1	0	1	
1	1	0	
1	1	1	

Figure 2-48

B. Assume that when all three switches are zero, the lamp is off, $S_1 = 0$, $S_2 = 0$, $S_3 = 0$, $L = 0$.

C. From this point, if one, and only one, switch changes, the lamp changes states. If two switches change, the lamp remains unchanged. If three switches change, the lamp will change states.

Given these conditions, the output of the truth table can be determined. The complete truth table looks like Figure 2-49.

D. From the truth table, the sum-of-products switching function can be written:

$$L = \overline{S}_1 \, \overline{S}_2 \, S_3 + \overline{S}_1 \, S_2 \, \overline{S}_3 + S_1 \, \overline{S}_2 \, \overline{S}_3 + S_1 \, S_2 \, S_3$$

E. To determine if the switching function can be simplified draw an N-cube (Figure 2-50).

F. The N-cube has no adjacent vertices, therefore, the function cannot be simplified.

G. The logic circuit is as shown in Figure 2-51.

S_1	S_2	S_3	L
0	0	0	0
0	0	1	1
0	1	0	1
0	1	1	0
1	0	0	1
1	0	1	0
1	1	0	0
1	1	1	1

Figure 2-49

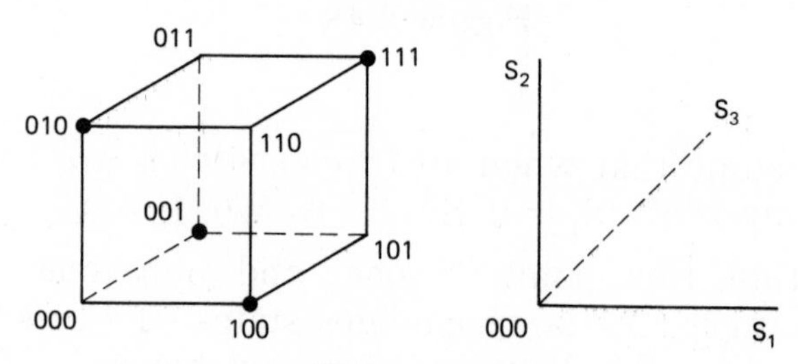

Figure 2-50

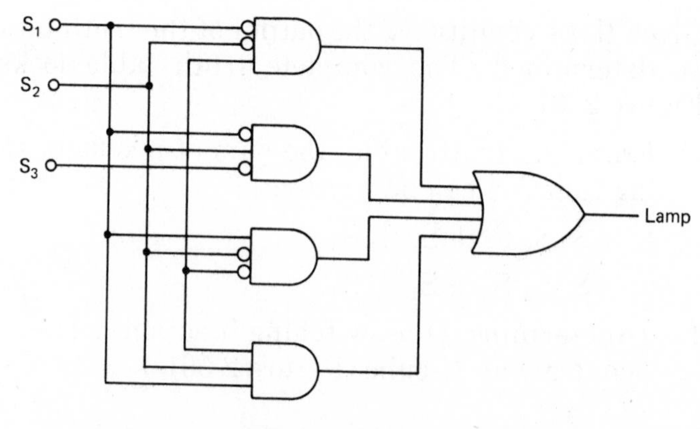

Figure 2-51

3 Definition of Logic Families

INTRODUCTION

AND gates, OR gates, and inverters are represented on logic diagrams as symbols. With their input/output relationships defined, you can follow through a logic circuit using these symbols and gain an understanding of these circuits. The gates themselves, however, are made up of individual components all built on a single silicon chip. Several families of logic circuits have been created and any given chip will be a member of a certain type of logic family. These families have their own characteristics in terms of switching speed, power consumption, noise immunity, and so on. This chapter gives examples of these families and outlines the differences between them.

EARLY LOGIC

Perhaps the first logic gates were made from switches. The basic AND and OR functions can be easily implemented with switches as seen in Figure 3-1.

Figure 3-1 A performs an AND function since both switches must be closed in order to light the lamp. The switches in Figure 3-1 B perform an OR function since the lamp will light with either or both of the switches closed.

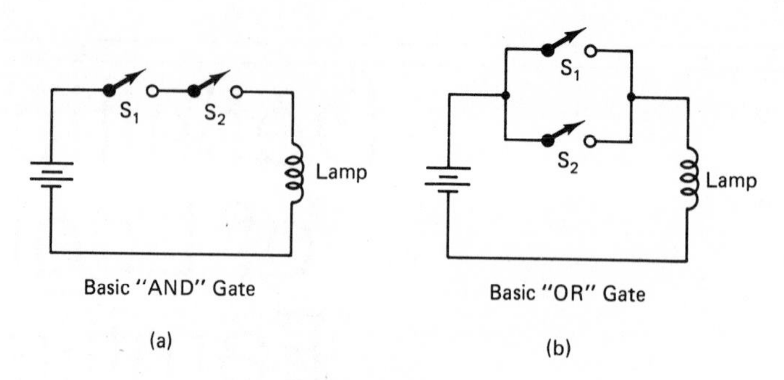

Figure 3-1

Diodes can also be used to build logic gates. The basic AND and OR functions are shown in Figure 3-2. The diode OR gate will produce a positive voltage at the output given a positive voltage at either input. The output voltage will be somewhat less than the input since there will be a voltage drop across the p-n junction. The diode AND gate will produce +V output when both inputs are +V or when both inputs are open. When one or both of the inputs are grounded the AND gate output will be close to zero. The actual output will be a small positive voltage because of the voltage drop across the p-n junction.

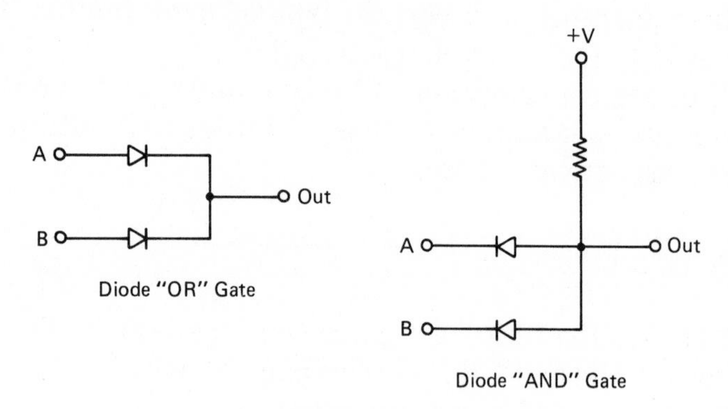

Figure 3-2

A simple inverter can be made from resistors and transistors. The circuit in Figure 3-3 will act as an inverter. Given a positive voltage on the input, the NPN transistor will conduct. This will lower the output voltage. In a properly designed circuit, the output voltage will drop to whatever the saturation voltage of the transistor is:

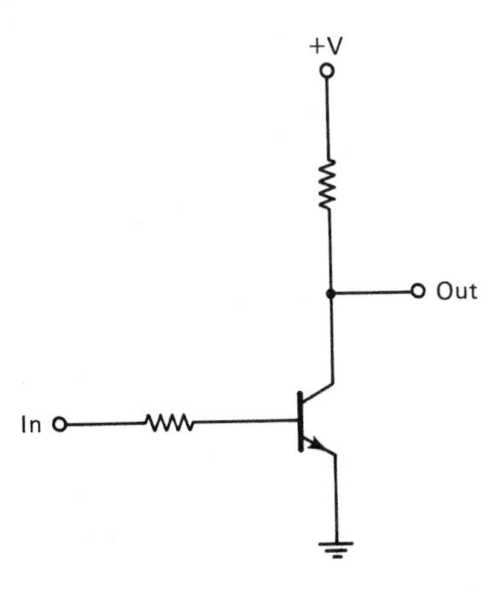

Figure 3-3

SWITCHING MODES

Logic gates constructed from bipolar transistors can be categorized into three main groups:

1. current sourcing
2. current sinking
3. current mode logic (CML)

1. *Current sourcing.* The circuit in Figure 3-4 shows the basic circuit using current sourcing. The base current needed to operate the second gate is derived from the previous gate. This is common operation in the logic family known as R.T.L. (resistor transistor logic).

2. *Current sinking.* The circuit in Figure 3-5 shows the basic circuit using current sinking. The base current for the gate is provided by a bias resistor. When circuit A is activated, it sinks the would-be base current, therefore changing the state of circuit B.

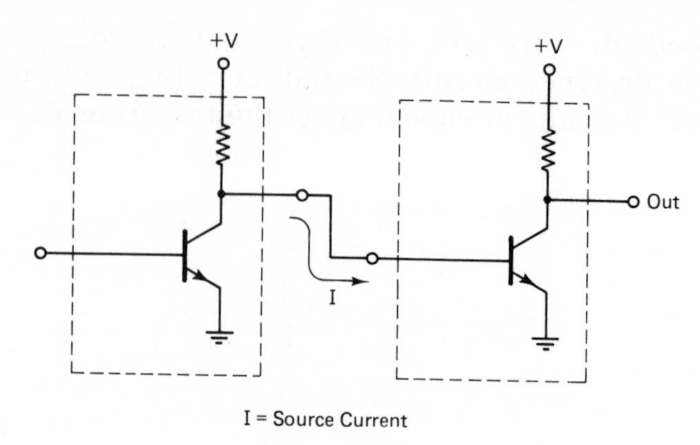

Figure 3-4

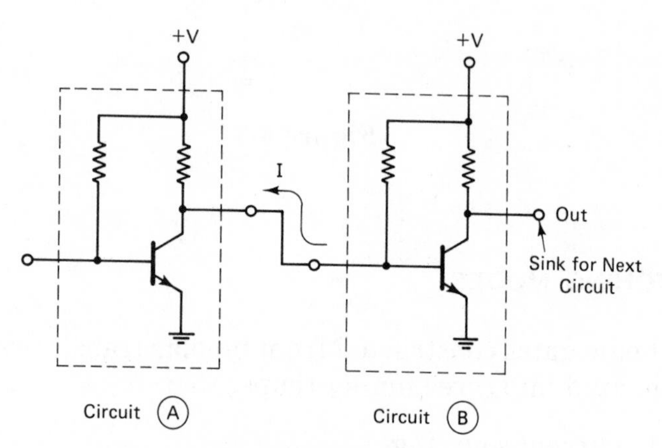

Figure 3-5

3. *Current mode logic.* The circuit in Figure 3-6 shows the basic circuit for current mode logic or current mode switching. The constant current source sends current through Q_2 since Q_1 is shut off with no input voltage. The output voltage in this case would be low since most of the supply voltage would be dropped across R. With a positive input voltage, Q_1 will now conduct. The fixed current that was flowing through Q_2 now flows through Q_1. The output voltage now goes high since there is no longer any voltage drop across R.

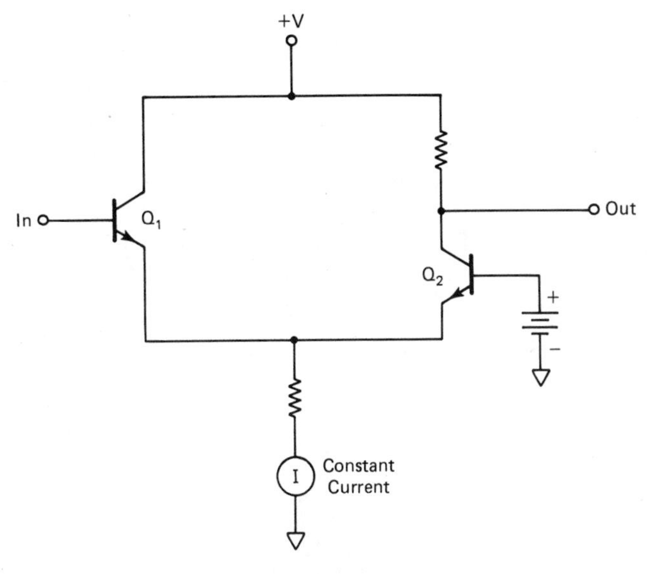

Figure 3-6

Current mode logic was devised in order to increase the frequency at which logic gates could operate. Other logic types operate the transistors in the saturation mode when conducting. Switching of the transistors is much slower when they are driven in and out of saturation.

RESISTOR TRANSISTOR LOGIC (RTL)

RTL logic is shown in Figure 3-7. It consists of transistors as the active switching element with resistors to limit the base current and as a voltage dropping device in the collector. A later version of this type of logic family is resistor, capacitor, transistor logic (RCTL) which is shown in Figure 3-8. If a capacitor is added in parallel to the input resistor, the maximum operating frequency of the gate is increased. When $R_1C_1 = R_2C_2$, the effect of the base-to-emitter capacitance is canceled. This effectively decreases the rise time of the input pulse and therefore increases the maximum operating frequency.

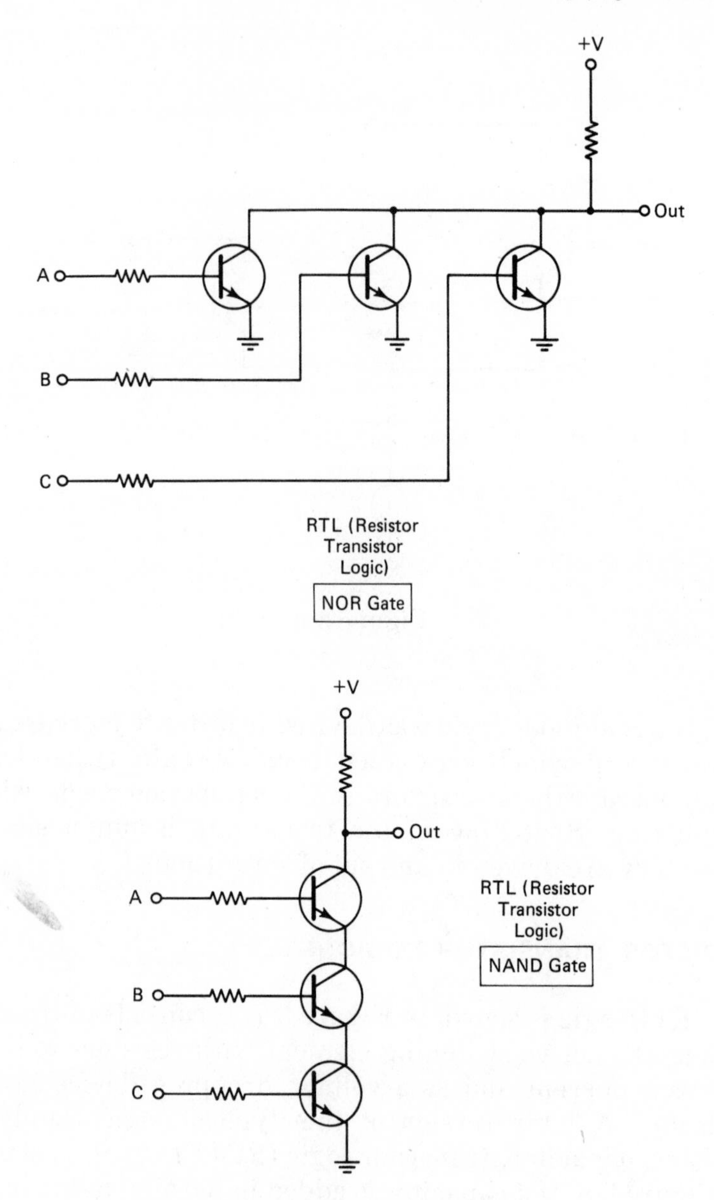

Figure 3-7: RTL (Resistor Transistor Logic).

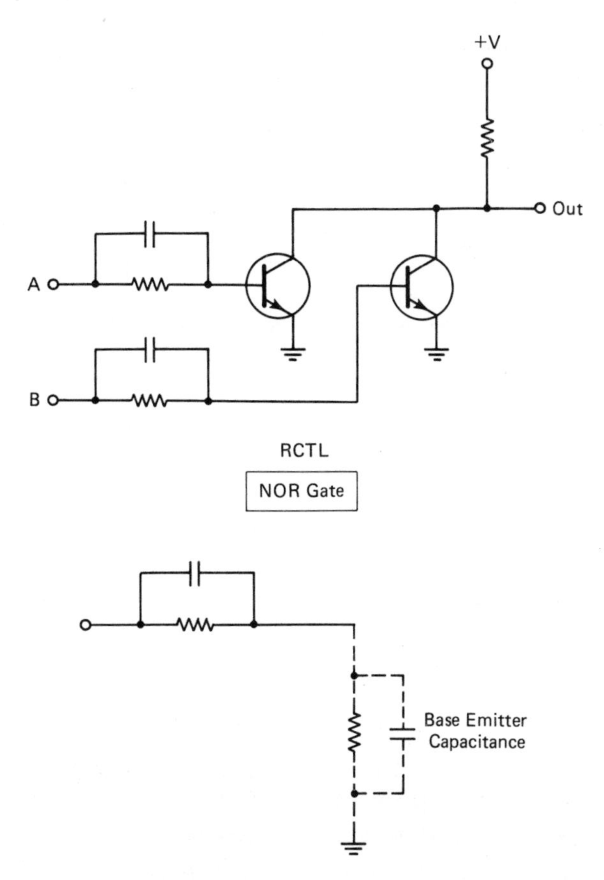

Figure 3-8: RCTL (Resistor Capacitor Transistor Logic). RCTL logic added a capacitor across the input resistor. This increased the operating frequency by compensating for the base/emitter capacitance.

DIODE TRANSISTOR LOGIC (DTL)

A DTL AND gate is shown in Figure 3-9. This circuit consists of a diode and gate followed by a transistor inverter. Diodes D_1 and D_2 provide a fixed voltage drop for the input to the transistor.

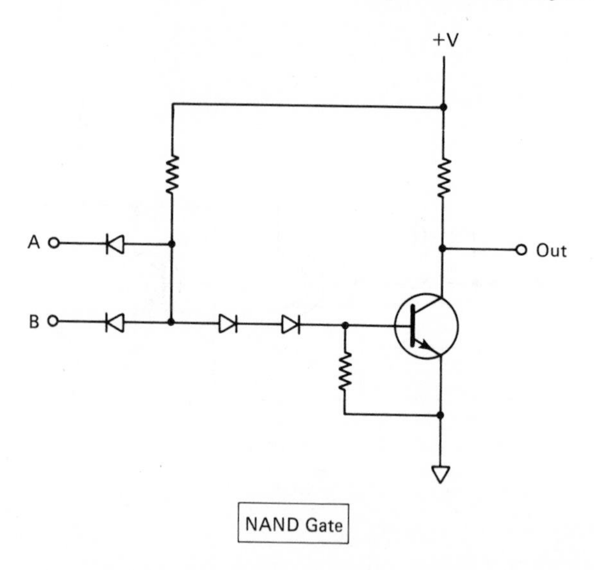

Figure 3-9: A DTL (Diode Transistor Logic) NAND Gate.

FAN-OUT

RTL, RCTL, and DTL logic gates typically operate from a supply voltage (Vcc) of 2.0 to 4.0 volts. The logic values of zero and one are represented by two different voltage levels. If the supply voltage is 3.5 volts, the ideal logic levels would be 3.5V equals 1 and 0.0V equals 0. In a real circuit, logic levels 1 and 0 are represented by a voltage range.

Logic 0 is usually a voltage value between 0.0 volts and 1.0 volts. When the transistor in Figure 3-10 is completely turned on (saturated), there is still a small voltage drop across the transistor. For a silicon transistor this voltage is approximately 0.7 volts.

When the transistor in Figure 3-10 is not conducting, the output voltage goes to 3.5 volts (Vcc). This is exactly true when the output is not connected to anything or is an open circuit. When the input of another gate is connected to this output, some value of current is drawn through R_1. This reduces the output voltage by the value (IR_1). If enough gates are connected to this output, the output voltage will

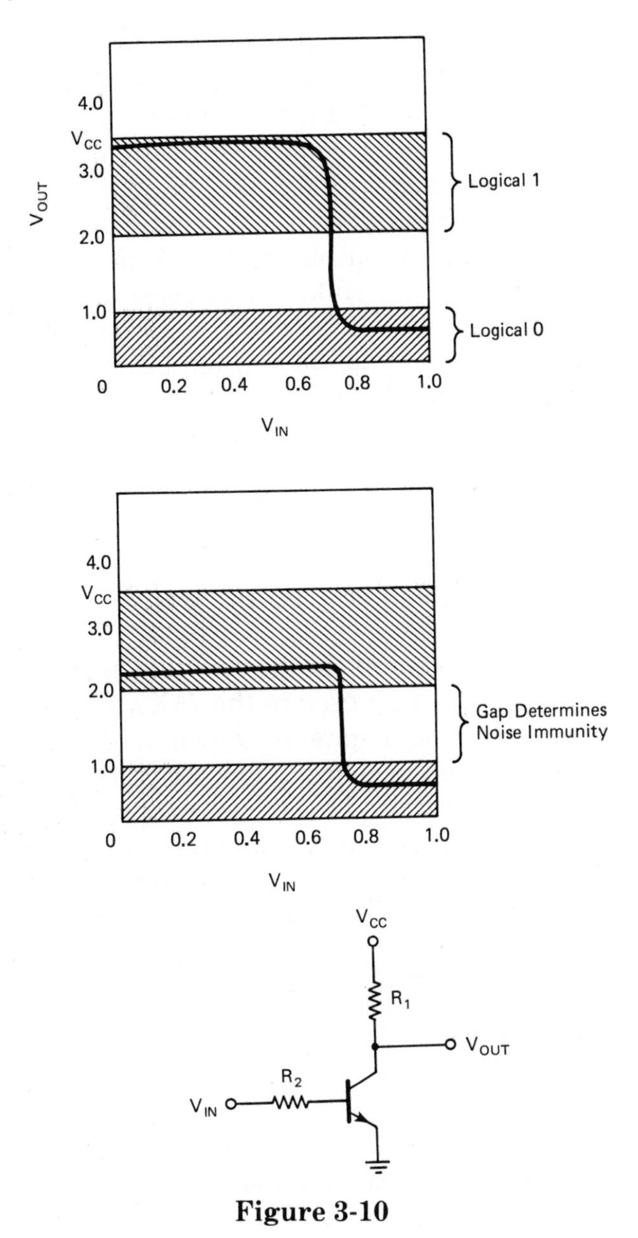

Figure 3-10

drop so much that it cannot act as a gate. The output voltage will be the same whether the transistor is conducting or not. It is common to define logical 1 as a range of positive voltage, say 2.0V to 3.5V as demonstrated in Figure 3-10. If enough gates are connected to the output to drop the output voltage below 2.0V, the gate will cease to function properly. The number of gates that it takes to do this is referred to as "fan-out." If five gates connected to the output causes the output voltage to drop to 2.0V, the "fan-out" is said to be five.

In the example shown in Figure 3-10, any voltage spike sufficient to raise the output from 0.7V to 2.0V can cause an error in the logic. This spike would cause a zero to look like a one. The amount of voltage spike necessary to cause this is determined by the "gap" in voltage between zero and one. The larger the gap, the better spike and noise immunity the logic circuit has. Noise immunity is also determined by the impedances designed into the circuit. The higher the impedance, the lower the noise immunity.

TRANSISTOR TRANSISTOR LOGIC (TTL)

TTL or T^2L is one of the most widely used logic families. TTL is the logic family used in the 74XX series of logic ICs. A typical TTL nand gate is shown in Figure 3-11. The family belongs in the current-sinking category defined earlier in this chapter, and all of the components used in the gates are made by doping a single piece of silicon.

The 7400 chip consists of TTL logic. It has four two-input NAND gates like the one in Figure 3-11, all integrated on one chip.

SCHOTTKY (T^2L)

In order to increase the operating frequency of T^2L type gates, a method to prevent the transistors from saturating was needed. The method used in Schottky T^2L logic was to place a Schottky diode between the base and collector of each transistor. A Schottky diode consists of a junction made from semiconductor and metal (aluminum). The diode prevents saturation by conduction of excess bias current. A Schottky T^2L NAND gate is shown in Figure 3-12.

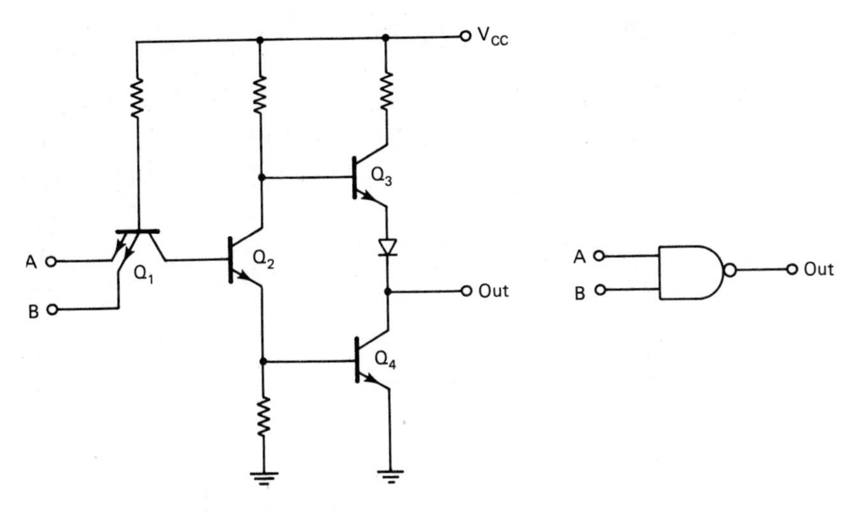

Figure 3-11: A Standard T²L NAND Gate.

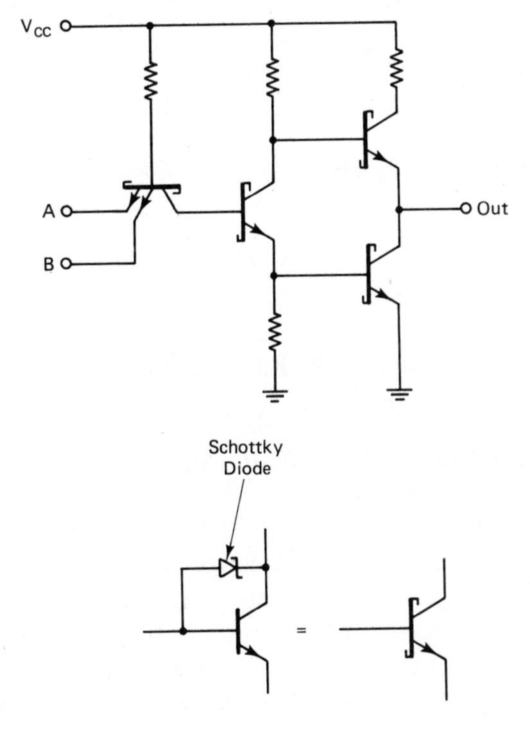

Figure 3-12: Schottky T²L.

EMITTER-COUPLED LOGIC (ECL)

Another method of getting around driving the transistors into saturation is to use the current switching mode described earlier in this chapter. Emitter-coupled logic employs current switching and emitter-follower amplifiers as shown in Figure 3-13 to perform logic functions. The circuit of Figure 3-13 has two inputs with both NOR and OR outputs.

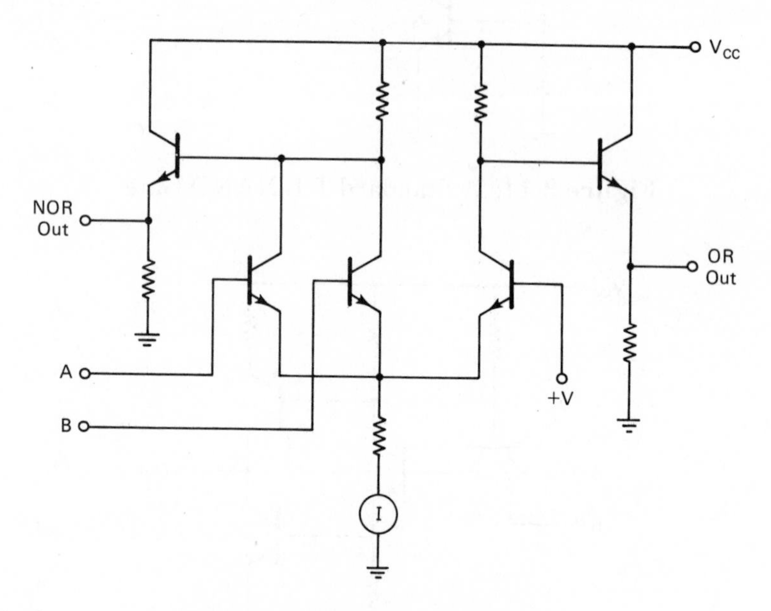

Figure 3-13: E.C.L. Logic Gate.

INTEGRATED INJECTION LOGIC (I^2L)

Integrated injection logic is a bipolar transistor logic that is a derivative of direct-coupled transistor logic. It is also known as merged transistor logic (MTL) since it uses transistors with common elements (emitters, collectors, and bases). Because of the common elements, I^2L advantages are (1) high speed and (2) high packing density on the silicon chip.

The derivation of an I^2L gate is shown in Figure 3-14. Part A is direct-coupled transistor logic, note the points A,

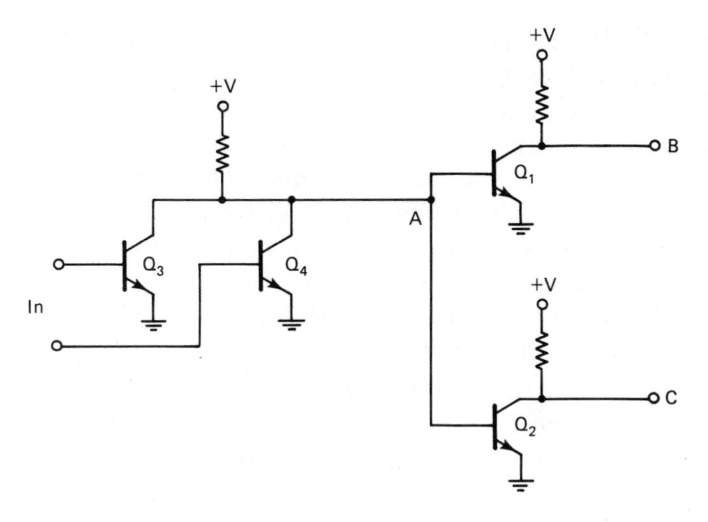

(a) DCTL-Direct-Coupled Transistor Logic

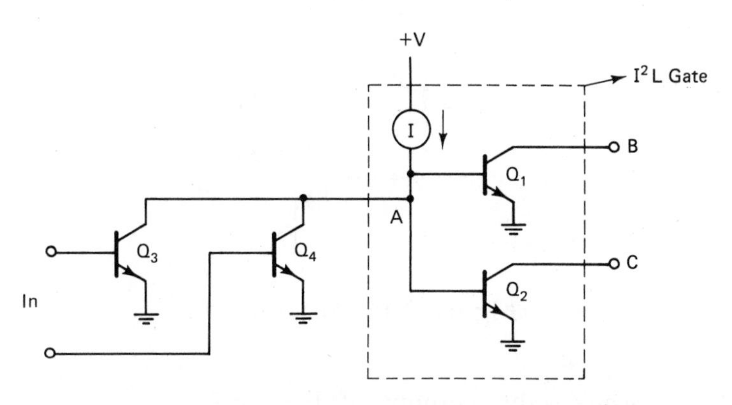

(b) DCTL Circuit Redrawn to I^2L Configuration

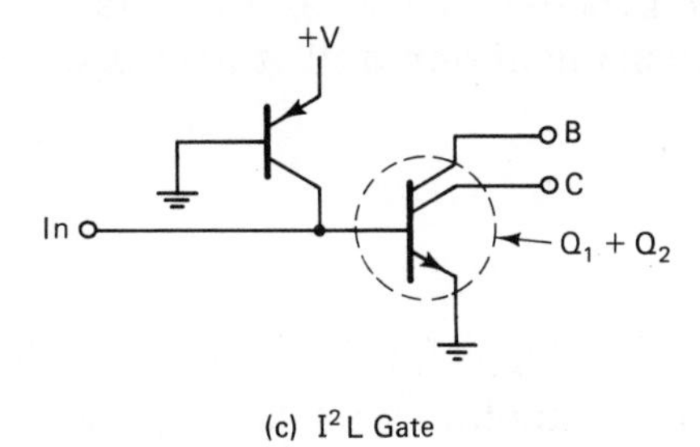

(c) I^2L Gate

Figure 3-14

B, and C. Part B shows the gate redrawn in I^2L configuration. Transistors Q_3 and Q_4 become current-sinking inputs for the I^2L gate shown in the dotted line. Transistors Q_1 and Q_2 can be combined into one transistor with two collectors since their bases are tied together. The constant current shown in Figure 3-14B is supplied by the PNP transistor shown in Figure 3-14C. The common elements in Figure 3-14C involve the PNP and NPN transistors. The collector of the PNP is the base of the NPN and the base of the PNP is the emitter of the NPN.

The standard I^2L gate can be enhanced by a couple of techniques. One technique is to use Schottky diodes in an analogous manner to the way they are used in T^2L gates. A Schottky I^2L gate is shown in Figure 3-15.

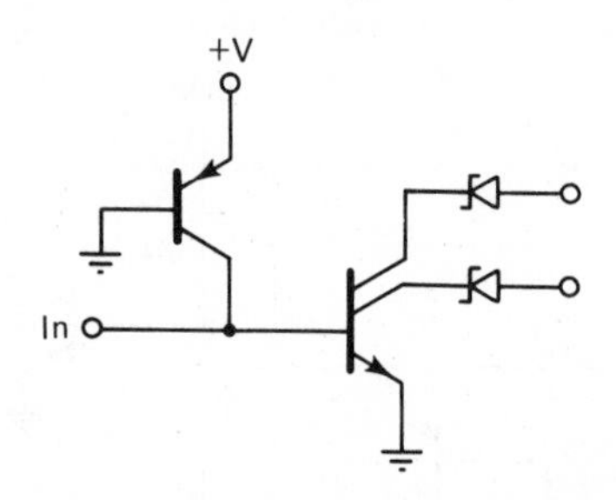

Figure 3-15: Schottky I^2L.

Another enhancement of I^2L logic is isoplanar integrated injection logic (I^3L). I^3L is I^2L manufactured using the isoplanar process. It uses oxide guard rings to prevent parasitic transistor action and it also uses ion implantation techniques.

MOS LOGIC

The previous logic families used, as the active switching device, the bipolar transistor. There exists another group of logic families made from metal-oxide-semiconductor field-effect transistor (MOSFETS). MOSFETS can take four basic forms:

1. N-channel, depletion mode
2. N-channel, enhancement mode
3. P-channel, depletion mode
4. P-channel, enhancement mode

The basic construction of a MOSFET is shown in Figure 3-16. The current path is between the source and drain. The source and drain are both heavily doped areas, N-for-N-channel, and P-for-P-channel. The current channel between source and drain is controlled by the metal gate. With the proper voltage on the gate, the device can be turned off or turned on. The gate is separated from the channel by a thin layer of silicon dioxide (S_iO_2). Silicon dioxide is the same as glass and this thin layer provides electrical insulation between the gate and the conducting channel. Because of this insulation, the input impedance of a MOSFET is extremely high.

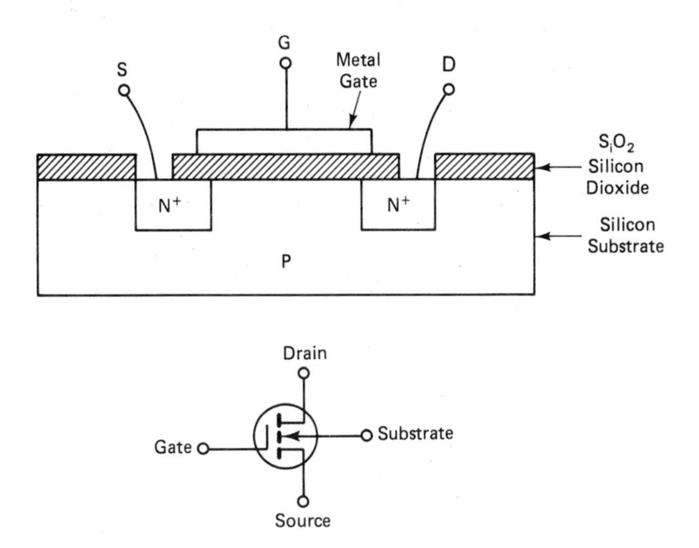

Figure 3-16: An enhancement mode, N-channel MOSFET. The arrow direction on the substrate denotes N-channel versus P-channel.

The basic difference between enhancement mode and depletion mode is shown in Figure 3-17. The depletion mode device has a conduction channel between source and drain doped into the substrate. It will conduct with zero volts applied to the gate. The enhancement mode device has no channel doped into the substrate. The channel is "enhanced" by a positive voltage applied to the gate. The enhancement mode device does not conduct with zero volts applied to the gate. It is for this reason that the enhancement mode device is the one most often used in logic gates.

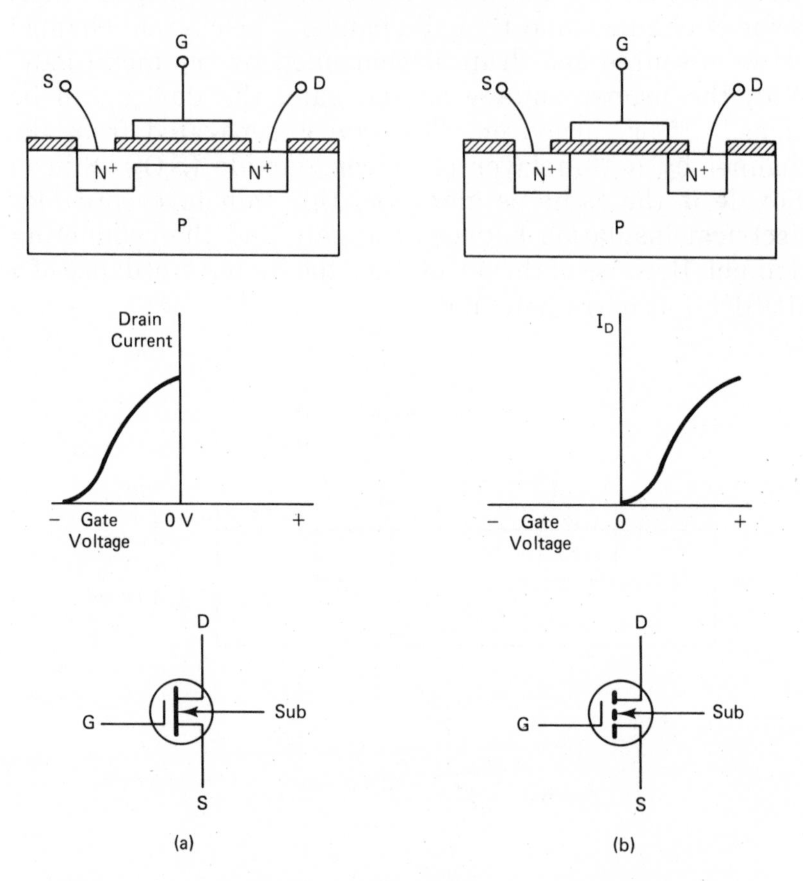

Figure 3-17: Enhancement mode (a) versus depletion mode (b) MOSFETS.

NMOS

The term NMOS applies to logic gates which use N-channel MOSFETS as the primary switching device. An example of NMOS logic is shown in Figure 3-18. This inverter consists of enhancement mode (N-channel) MOSFETS. With zero volts on the input, Q_2 is turned off and Q_1 is turned on. The output is high and is equal to the supply voltage minus the drop across Q_1. With a positive voltage on the input (Binary 1), Q_2 conducts and the output voltage drops to near zero.

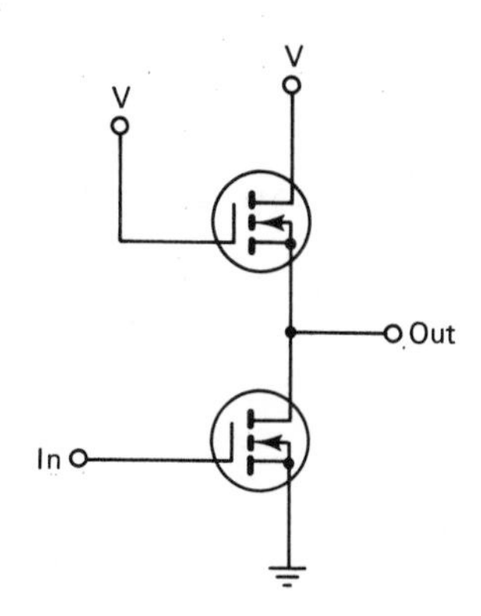

Figure 3-18: NMOS Inverter.

PMOS

The term PMOS applies to logic gates which use P-channel MOSFETS as the primary switching device. An example of a PMOS inverter is shown in Figure 3-19. Its action is the same as the NMOS device except that the P-channel works on the opposite polarity.

VMOS

Another type of MOSFET technology involves the VMOS or vertical MOS. The advantage of VMOS is its power-handling capability which is several times greater than NMOS or PMOS.

The term "vertical" comes from the manner in which the device is fabricated. Figure 3-20 shows the cross section of a vertical MOSFET with an N-channel. The substrate becomes the drain in this device. The N-layer between the

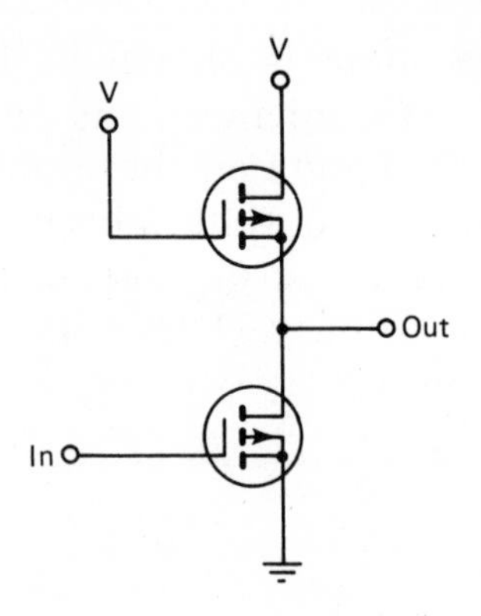

Figure 3-19: PMOS Inverter.

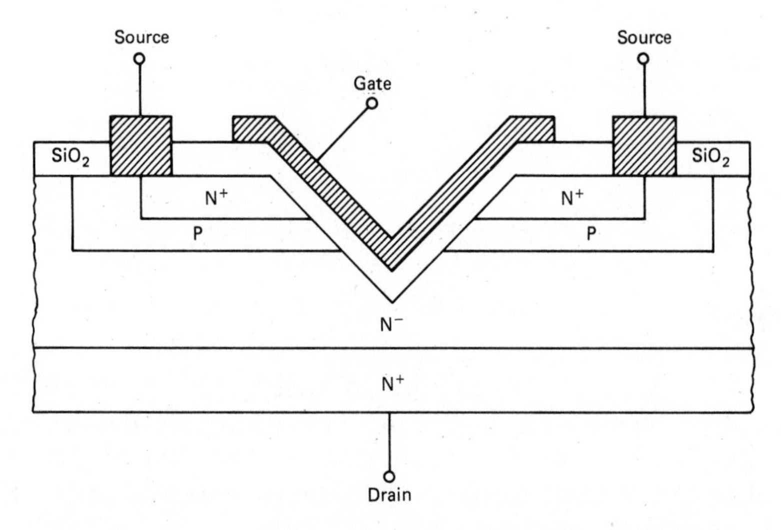

Figure 3-20: Cross Section of a Vertical MOS Transistor.

source and drain is there to absorb the junction depletion region and, therefore, increases the drain-source breakdown voltage.

CMOS

CMOS stands for "complementary" MOS. The word complementary comes from the fact that CMOS uses both N-channel and P-channel MOS devices on the same silicon chip. A CMOS inverter is shown in Figure 3-21. When the input voltage is low (binary 0) the P-channel device is turned on and the N-channel device is turned off. This gives a high (binary 1) output. When the input voltage goes high (binary 1) and P-channel device turns off and the N-channel device turns on. This gives a very low output voltage (binary 0). An example of a CMOS NAND gate is shown in Figure 3-22.

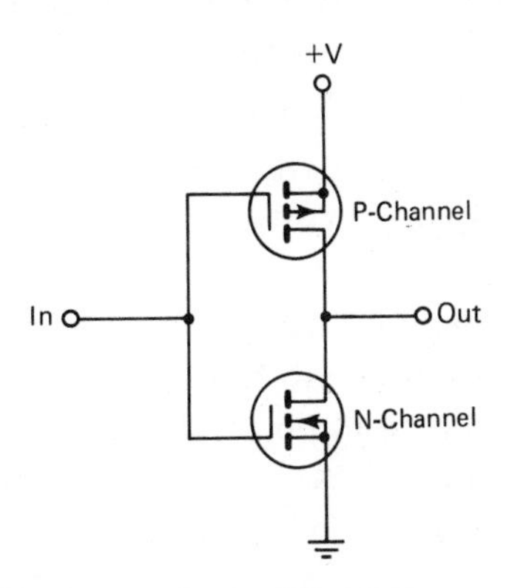

Figure 3-21: A CMOS Inverter.

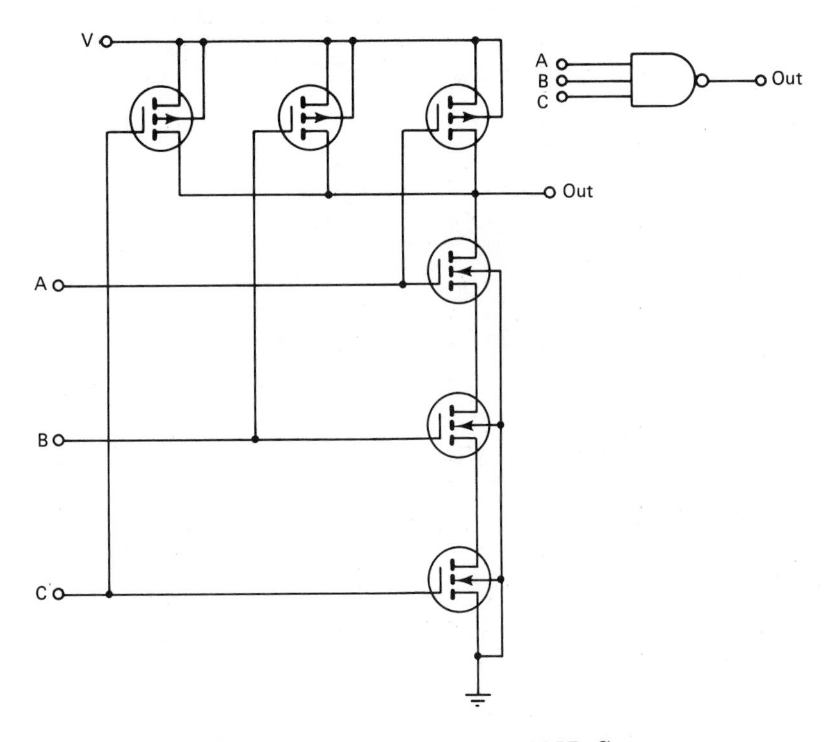

Figure 3-22: A CMOS NAND Gate.

CMOS logic circuits have become quite common and are available in "off-the-shelf" chips. They have both advantages and disadvantages; some of which are listed below:

A. To generate an N-channel or P-channel MOSFET on a silicon chip, only one diffusion step is necessary. To generate both type devices on the same chip requires twice as many diffusion steps.

B. The CMOS logic gate not only has extremely high input impedance, it has very little current drain from the supply voltage. From the CMOS inverter example in Figure 3-21, whenever one device is turned on the complementary device is turned off. Except for current transitions during the switching time, there is no current drain from the supply voltage. In applications where low power consumption is a must, CMOS logic is almost always used.

C. Because of the high input impedances, CMOS logic is highly susceptible to noise and electrical static discharges. Errors can, and do, occur because of this.

4

Checklist of Logic Circuits

INTRODUCTION

This chapter contains examples of logic circuits used in the arithmetic operation of numbers. The first section shows circuits that perform addition and subtraction on binary numbers. The second section shows circuits used to convert numbers from one base system to another and from various coding schemes to other number systems.

ARITHMETIC CIRCUITS

Logic circuits that perform binary arithmetic all start from circuits that add two binary digits. Addition of two digits is shown in Figure 4-1. The definition of mod-2 addition, the Boolean algebra switching functions, as well as several logic circuits are all shown in Figure 4-1. The half-adder or exclusive-OR circuit can add two binary digits and produce a "carry" ouput if the circuit is chosen properly. What the half-adder cannot do is add three binary digits.

Full-Adder

A full-adder consists of two half-adders connected in series as shown in Figure 4-2. The output of one half-adder is fed to the input of the second half-adder along with the carry-in. This circuit is capable of adding three binary digits and is used in several ways to add numbers.

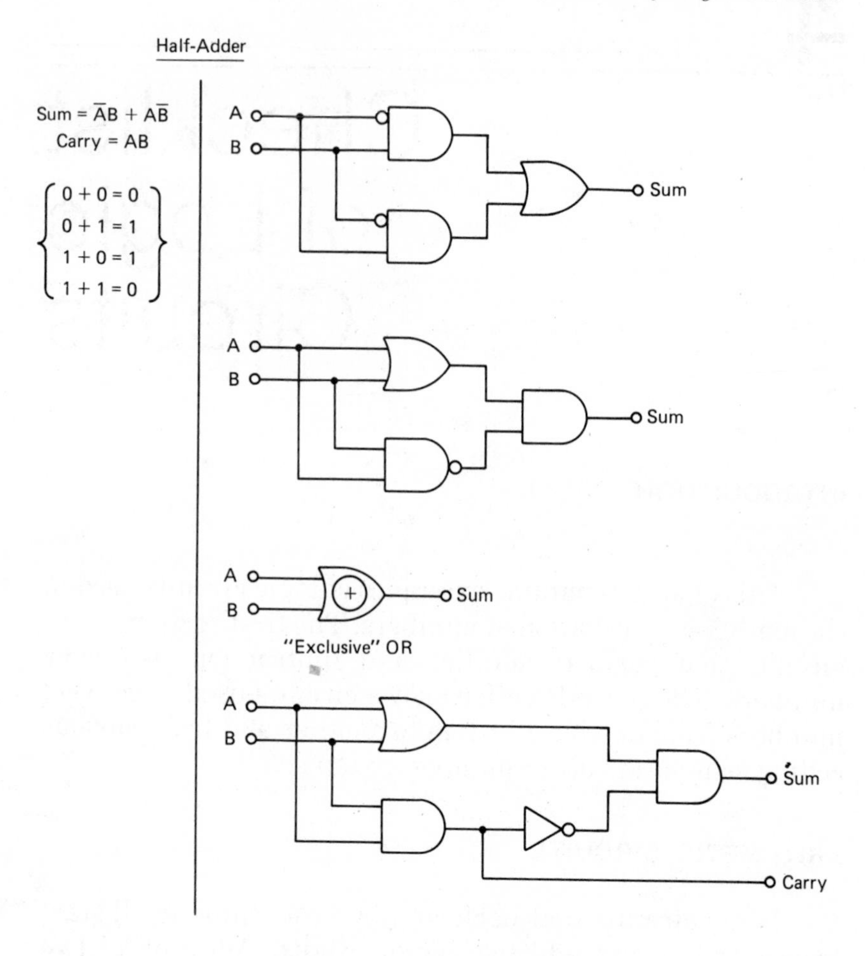

Figure 4-1: Half-Adder.

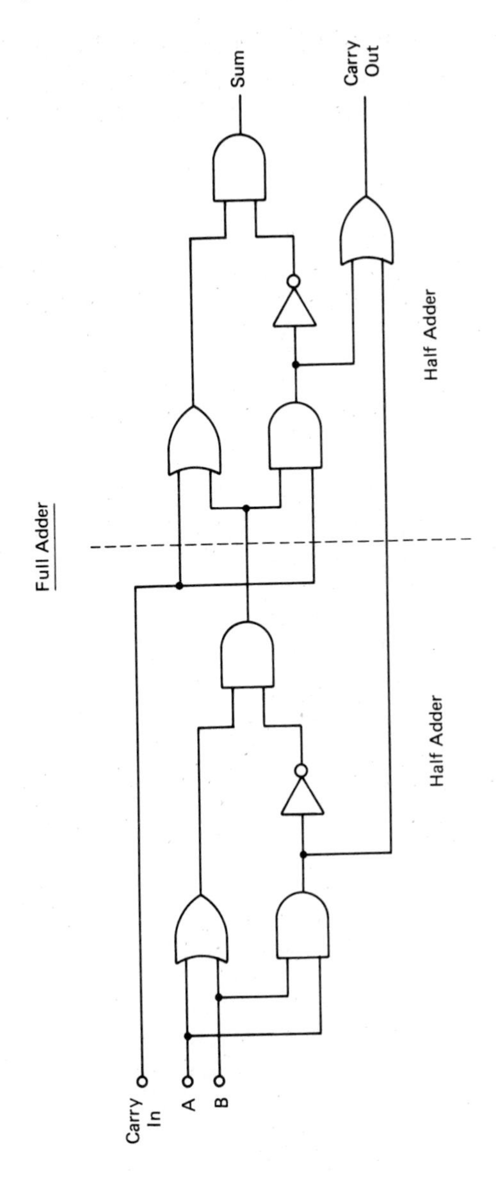

The Full-Adder Consists of Two Half-Adders with the Output of One Fed to the Input of the Next. There will be a Carry Out of the Full-Adder Whenever Either Half-Adder has a Carry.

Figure 4-2: Full Adder.

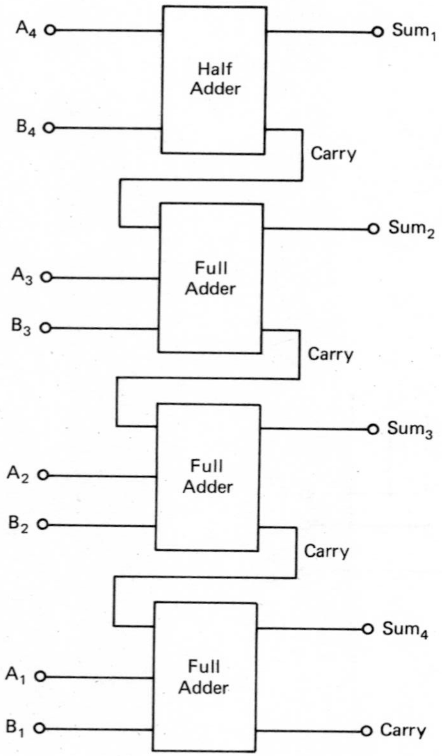

	A_1	A_2	A_3	A_4
+	B_1	B_2	B_3	B_4
Carry	Sum_4	Sum_3	Sum_2	Sum_1

Four-Bit Binary Numbers can be Added Using the Circuit of Figure (4-3). Two Four-Bit Numbers will Add to Produce a Five-Bit Binary Number.

The Addition of the Least Significant Bits (A_4, B_4) Requires Only a Half-Adder Since There Cannot be any Carry In.

			Carry = 1
For:	$A_1 = 1$	$B_1 = 1$	$Sum_4 = 0$ (with Carry)
	$A_2 = 0$	$B_2 = 0$	$Sum_3 = 1$
	$A_3 = 1$	$B_3 = 1$	$Sum_2 = 0$ (with Carry)
	$A_4 = 0$	$B_4 = 1$	$Sum_1 = 1$

Figure 4-3: Parallel Full Adders.

One use of the full-adder is shown in Figure 4-3. This circuit uses full-adders to add numbers in parallel. The circuit used for the least significant bits is a half-adder since there cannot be a carry input at this stage. The rest of the circuits are full-adders. There is one full-adder for each binary position. An example of the addition of two four-bit binary numbers is also given in the example. Any four-bit binary numbers can be added using this circuit. If eight-bit binary numbers were to be added, four more full-adders could be added to the circuit.

Full-adders can also be used to add binary numbers in serial form. The circuit shown in Figure 4-4 shows the basic idea. The binary numbers are contained in two shift registers. The numbers are shifted through the full-adder two digits at a time. The sum is shifted into a third register at the same time. The carry function is handled by a flip-flop that delays the carry so that it can be added along with the next two binary bits. Figure 4-5 shows the addition of two binary numbers using the circuit of Figure 4-4. The on/off status of each flip-flop is shown versus the clock pulses which control the circuit.

Half-Subtractor

A logic circuit for a half-subtractor is shown in Figure 4-6. The difference output is the same as the sum output of a half-adder and therefore the switching function is also the same. The only change between the two is in the carry and borrow functions. The borrow function of a half-subtractor is borrow AB.

Full-Subtractor

Analogous to the half-adder and full-adder, two half-subtractors can be connected to form a full-subtractor. A full-subtractor is shown in Figure 4-7.

Full-subtractors can be connected in order to subtract binary numbers as shown in Figure 4-8(A). It is more common to use the same circuit to both add and subtract binary

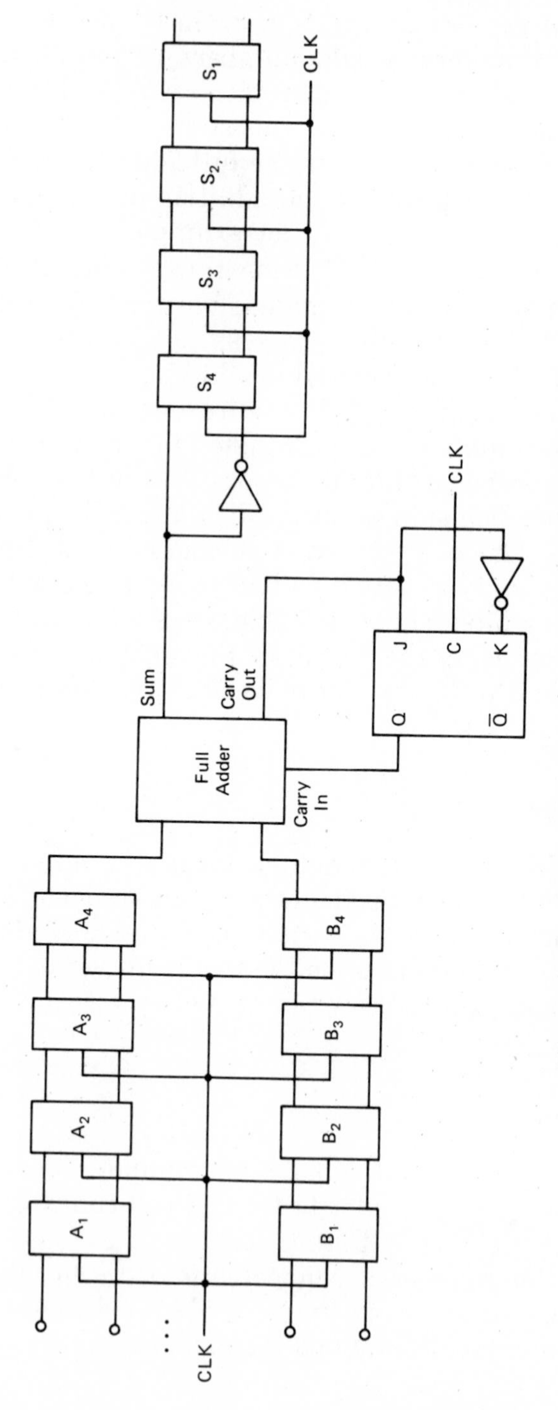

Figure 4-4: Full Adders Used to Add Binary Numbers in Serial Form.

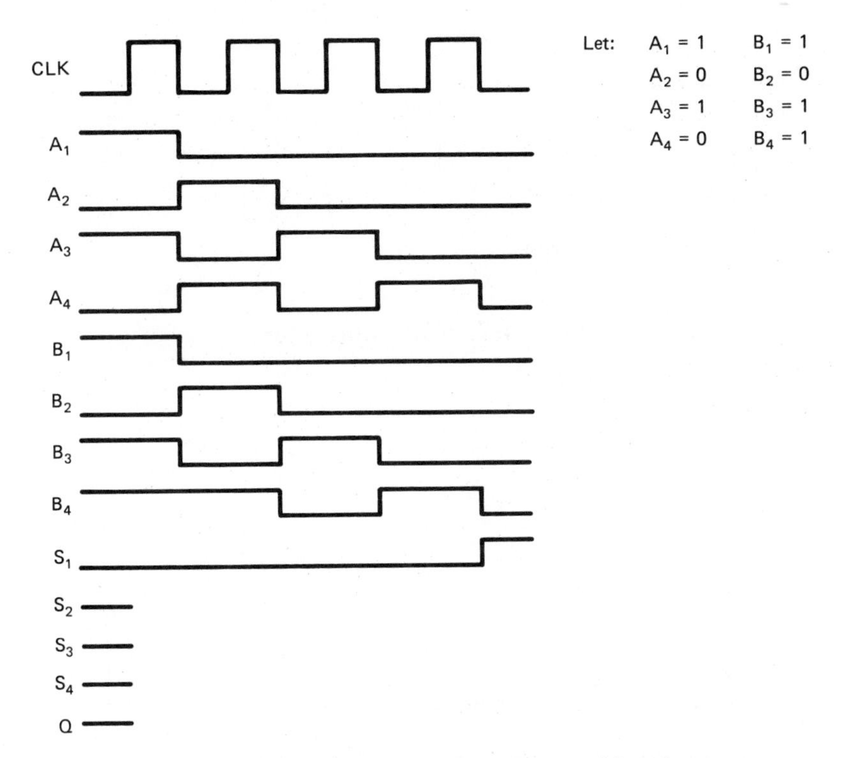

Figure 4-5: Addition of Two Binary Numbers through the Circuit in Figure 4-4.

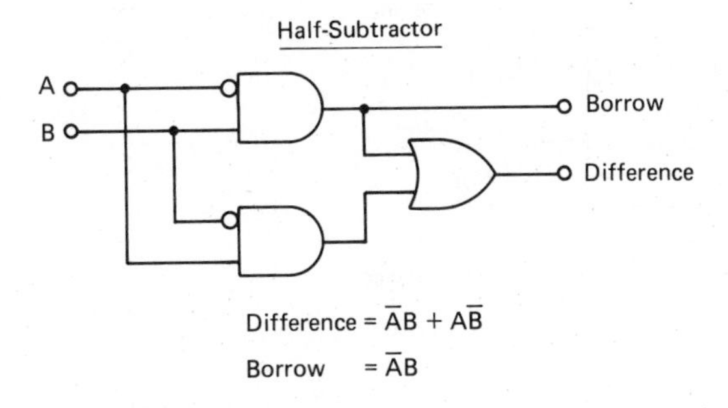

Figure 4-6: Half-Subtractor.

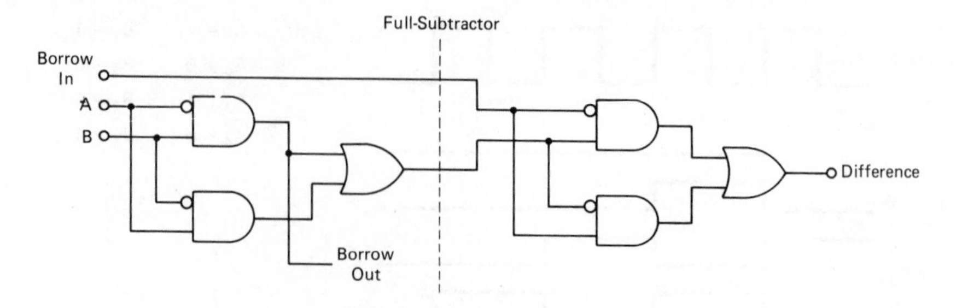

Figure 4-7: Full Subtractor.

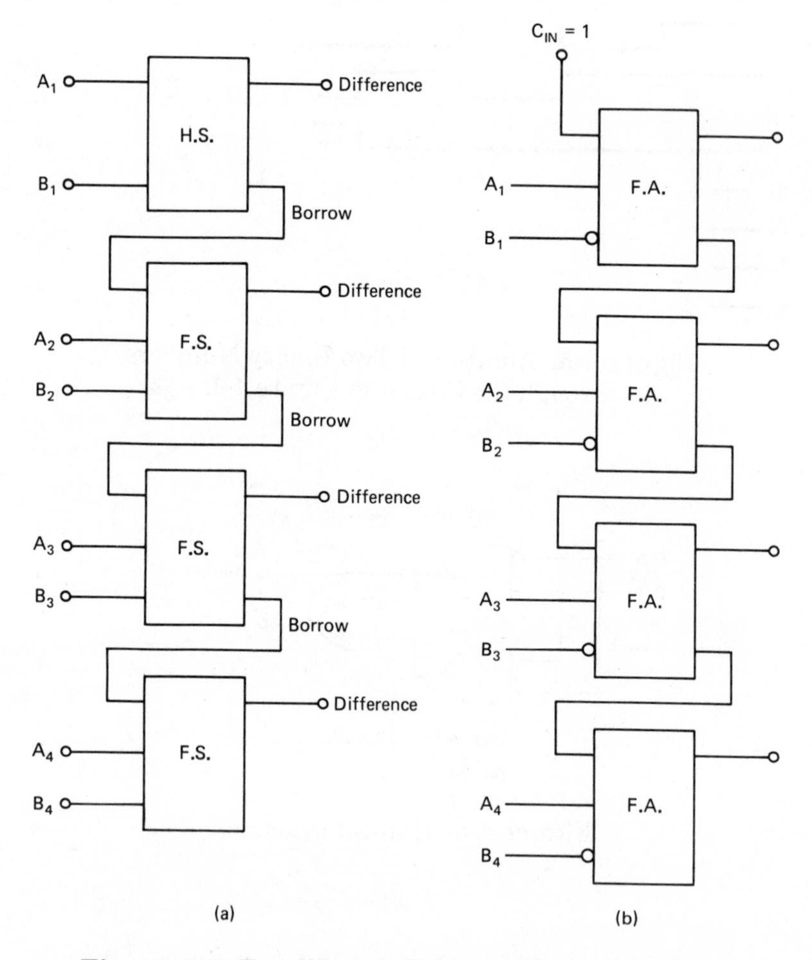

Figure 4-8: Two Ways to Subtract Binary Numbers.

numbers. It is possible to subtract binary numbers by using complements and adding as discussed in Chapter #1. Figure 4-8(B) shows how to use the two's complement to subtract. All of the B inputs are inverted prior to each full-adder. This combined with the one inputted at the first carry input creates the two's complement of the binary number to be subtracted. Subtraction is then accomplished by adding the two's complement.

NUMBER SYSTEM DECODERS

It is often necessary to convert number systems from one base to another. This may take the form of conversion from one base system to another base system, or it may take the form of coded numbers to uncoded numbers. Certain conversions are common enough so that separate ICs have been designed just to perform the conversion.

Several of the common conversions along with the logic circuits needed to implement them are discussed in the following sections.

Two-Variable Decoder

A two-variable decoder or a 2-of-4 decoder is shown in Figure 4-9. This circuit takes the four possible combinations of two inputs and makes one and only one output high for each input. The input/output relationship is shown in the table in Figure 4-9. Y_0 is the high output for the $(A = O_1, B = O)$ input, Y_1 is the high output for $(A = O, B = 1)$ etc. One of the four outputs is high for any combination of inputs, hence the name "two variable decoder."

Three-Variable Decoder

Given three inputs, there exists $2^3 = 8$ possible combinations of inputs. The purpose of a three-variable decoder is to convert each of these input possibilities into a single output. The three-variable or 3-of-8 decoder is shown in Figure 4-10. The table shows that for each different input combination, only one of the eight outputs is a binary one.

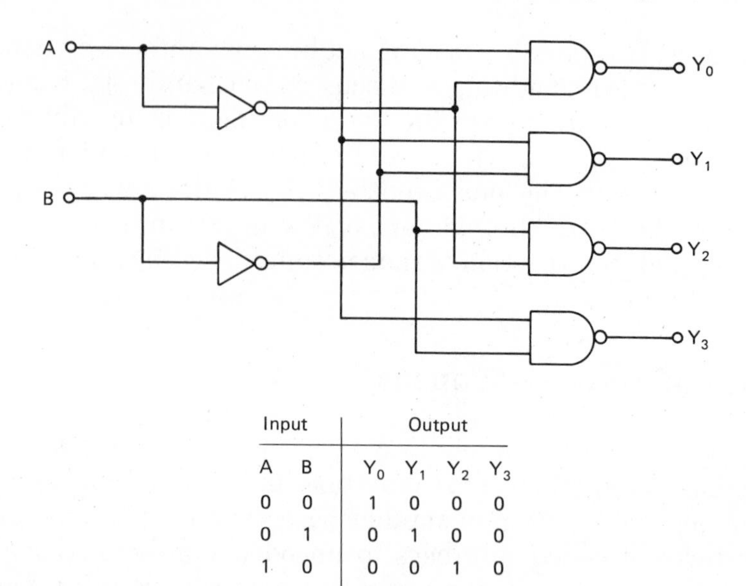

Input		Output			
A	B	Y_0	Y_1	Y_2	Y_3
0	0	1	0	0	0
0	1	0	1	0	0
1	0	0	0	1	0
1	1	0	0	0	1

Figure 4-9: Two-Variable Decoder (2 of 4).

BCD to Decimal

Binary coded decimal (BCD) consists of four binary digits representing the decimal digits zero through nine. The BCD-to-decimal converter has four inputs and converts each combination of these inputs into a single output. The circuit is shown in Figure 4-11 along with the input/output table for the circuit.

Binary to Octal

One way to convert a binary number into an octal number is to group the binary digits into groups of three as shown in Chapter 1. Each group of three binary digits forms one octal digit.

To convert to octal, each binary group of three is fed into a 3-of-8 converter. The output of the 3-of-8 converter forms the octal digits. This system is shown in Figure 4-12.

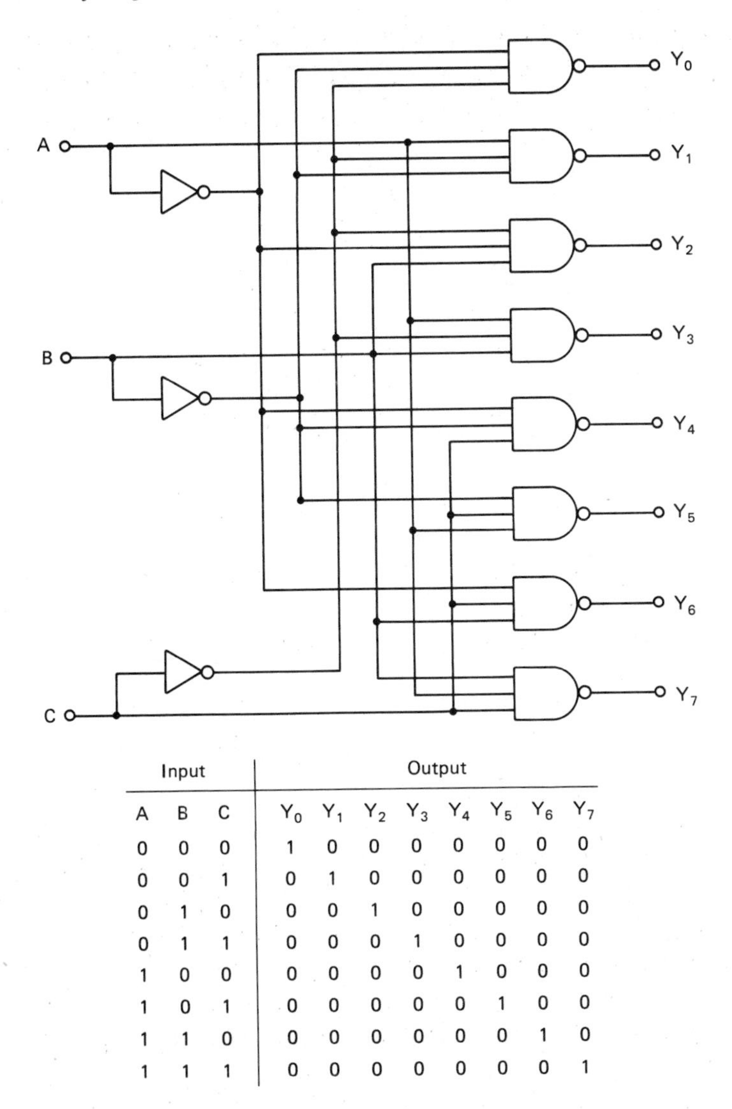

Input			Output							
A	B	C	Y_0	Y_1	Y_2	Y_3	Y_4	Y_5	Y_6	Y_7
0	0	0	1	0	0	0	0	0	0	0
0	0	1	0	1	0	0	0	0	0	0
0	1	0	0	0	1	0	0	0	0	0
0	1	1	0	0	0	1	0	0	0	0
1	0	0	0	0	0	0	1	0	0	0
1	0	1	0	0	0	0	0	1	0	0
1	1	0	0	0	0	0	0	0	1	0
1	1	1	0	0	0	0	0	0	0	1

Figure 4-10: Three-Variable Decoder (3 of 8).

Excess 3 to Decimal

The circuit used to decode excess 3 binary numbers to decimal is shown in Figure 4-13. The circuit is quite similar to the BCD-to-decimal converter except for the "excess three" added to the binary number. The input/output relationship is also shown in Figure 4-13.

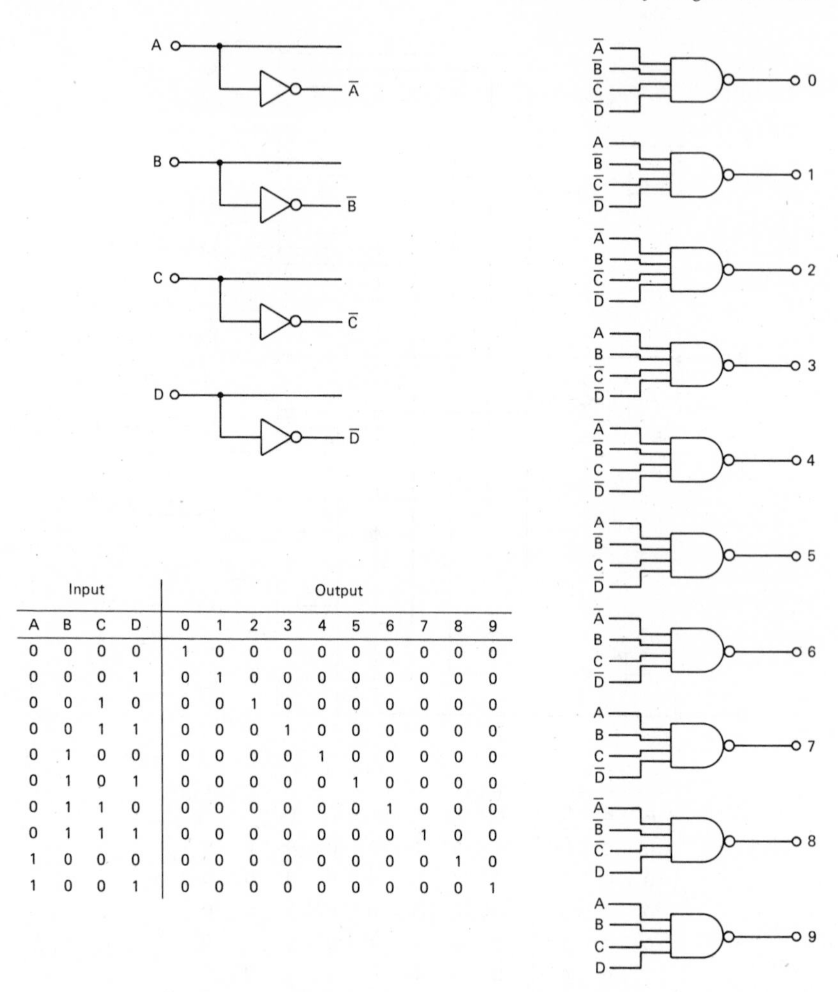

Input				Output									
A	B	C	D	0	1	2	3	4	5	6	7	8	9
0	0	0	0	1	0	0	0	0	0	0	0	0	0
0	0	0	1	0	1	0	0	0	0	0	0	0	0
0	0	1	0	0	0	1	0	0	0	0	0	0	0
0	0	1	1	0	0	0	1	0	0	0	0	0	0
0	1	0	0	0	0	0	0	1	0	0	0	0	0
0	1	0	1	0	0	0	0	0	1	0	0	0	0
0	1	1	0	0	0	0	0	0	0	1	0	0	0
0	1	1	1	0	0	0	0	0	0	0	1	0	0
1	0	0	0	0	0	0	0	0	0	0	0	1	0
1	0	0	1	0	0	0	0	0	0	0	0	0	1

Figure 4-11: BCD-to-Decimal Decoder.

BCD to Seven-Segment

LED displays are constructed in seven segments as shown in Figure 4-14. All of the decimal characters from zero to nine can be represented by these LEDs by illuminating the proper segments. A one is displayed by illuminating segments B and C. A nine is displayed by illuminating segments G, F, A, B, and C.

The purpose of the B.C.D.-to-seven-segment decoder is to take the binary output of a counter or other device and decode the data in order to illuminate the correct segments of the L.E.D. display. The logic necessary to do this is shown in Figure 4-14. All of this logic can be found integrated into single IC chips.

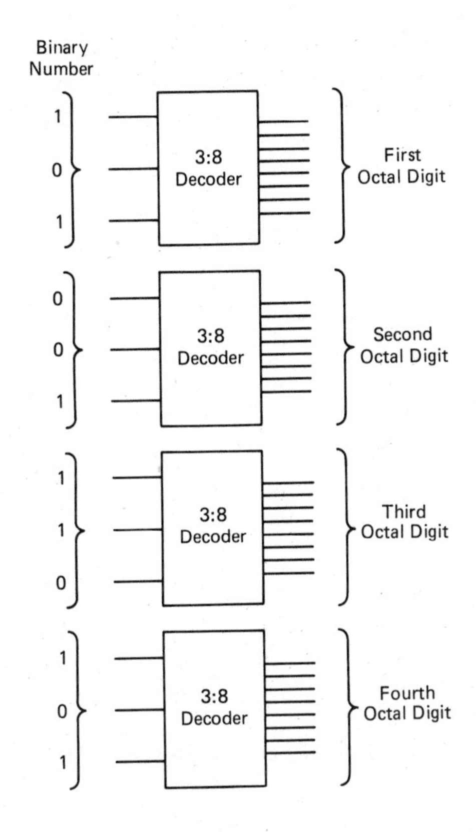

Figure 4-12: Converting Binary Numbers to Octal Numbers.

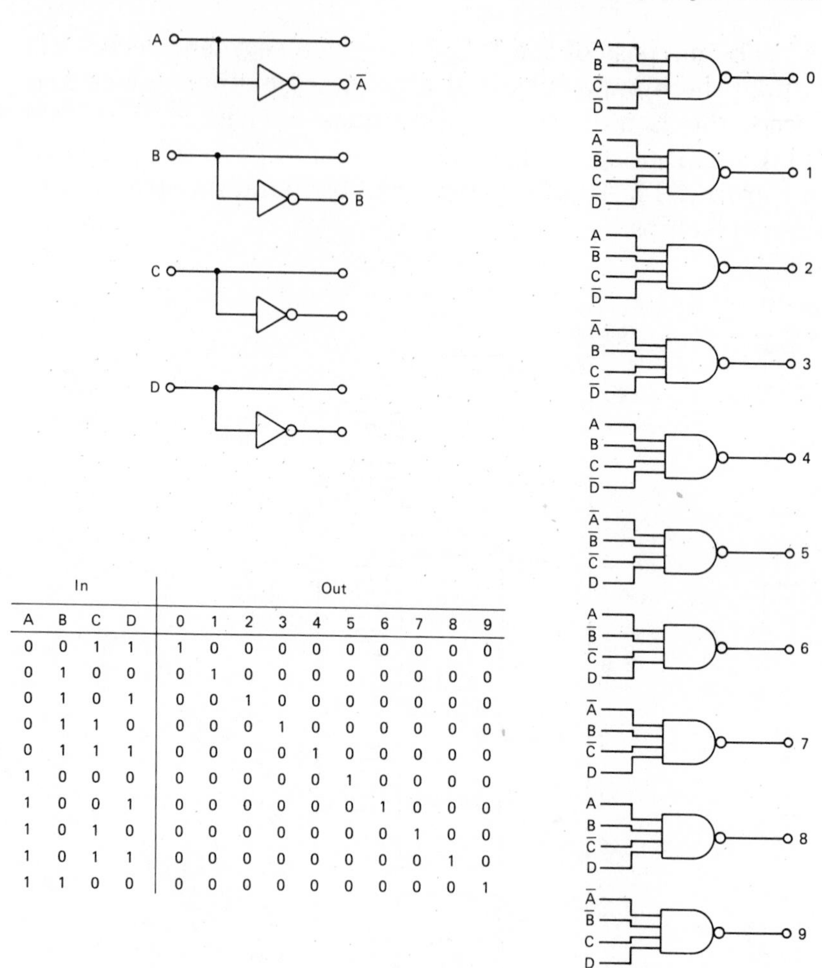

In				Out									
A	B	C	D	0	1	2	3	4	5	6	7	8	9
0	0	1	1	1	0	0	0	0	0	0	0	0	0
0	1	0	0	0	1	0	0	0	0	0	0	0	0
0	1	0	1	0	0	1	0	0	0	0	0	0	0
0	1	1	0	0	0	0	1	0	0	0	0	0	0
0	1	1	1	0	0	0	0	1	0	0	0	0	0
1	0	0	0	0	0	0	0	0	1	0	0	0	0
1	0	0	1	0	0	0	0	0	0	1	0	0	0
1	0	1	0	0	0	0	0	0	0	0	1	0	0
1	0	1	1	0	0	0	0	0	0	0	0	1	0
1	1	0	0	0	0	0	0	0	0	0	0	0	1

Figure 4-13: Excess 3-to-Decimal Decoder.

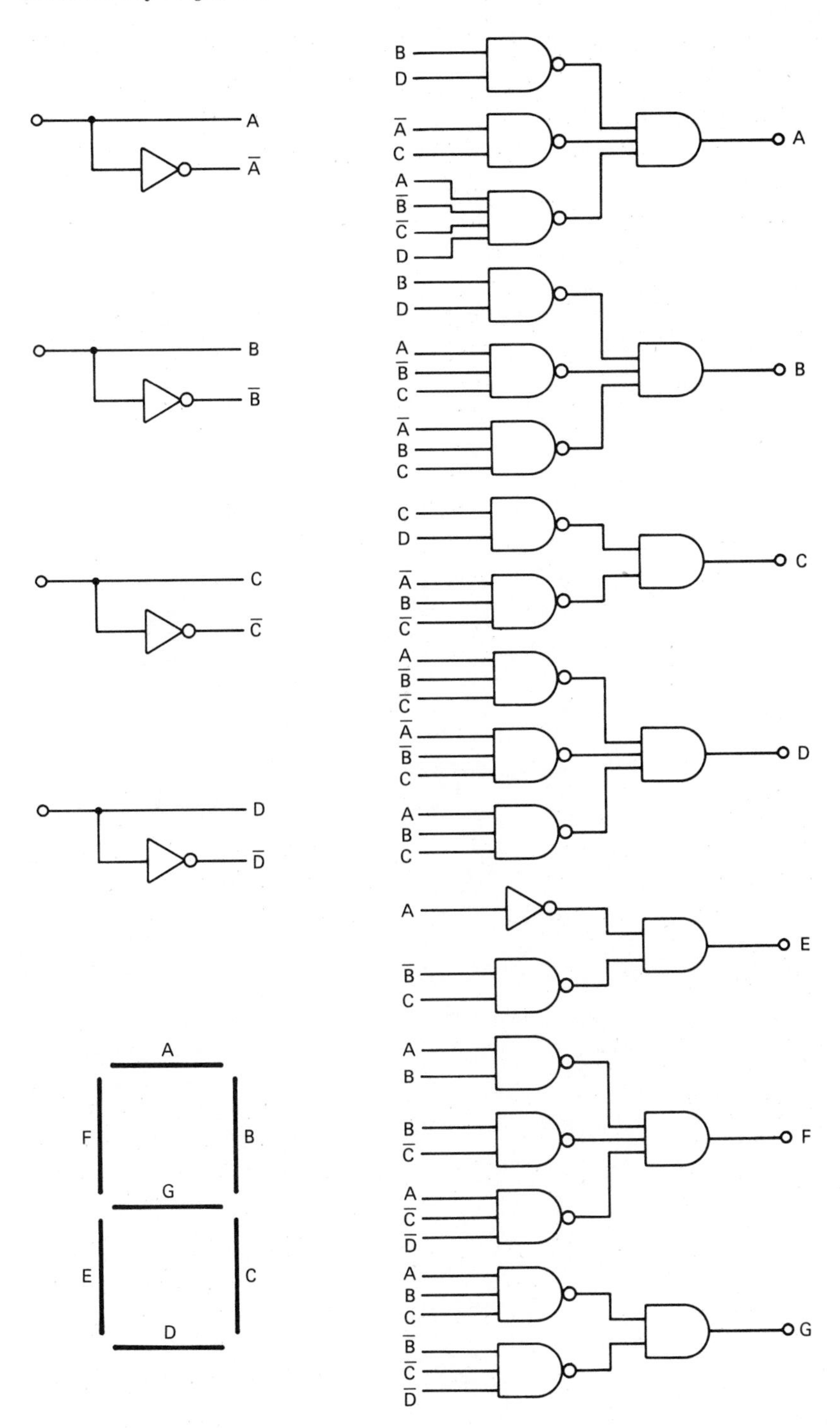

Figure 4-14: BCD-to-Seven-Segment Decoder.

5

Examples of Flip-Flops and Counters

INTRODUCTION

Sequential networks, as opposed to combinational networks, have their output dependent on a past sequence of inputs as well as present inputs. This ability to "remember" gives sequential networks their special place in digital logic. The basis of almost all sequential networks is the flip-flop. The flip-flop is a device with two stable states and has a memory ability since it will remain in a stable state until changed by new inputs.

THE LATCH

The latch is a flip-flop in its most basic form. It usually consists of NAND or NOR gates with a feedback loop as shown in Figure 5-1. When power is applied to this circuit it will settle into one of two stable states. To pick a state let $A = 1$ and $B = 0$. Since A is one, its input to N_2 keeps B at

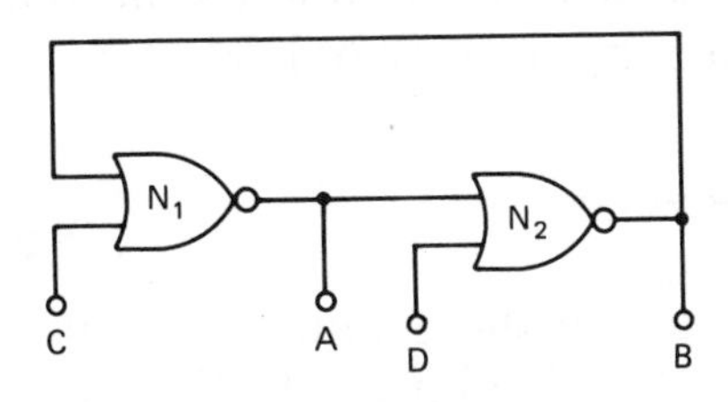

Figure 5-1

zero and this zero fed back to N_1 keeps A at one. The latch will remain in this state until an external input changes the state.

Again in Figure 5-1 let A = 1, B = 0 and now place a one at C. Since the input to N_1 is now 0, 1, the output will go to zero. This zero will force B to go to one. This one will be fed back to the input of N_1 and hold the output A at zero.

The input at C caused the latch to toggle, and A and B changed states. The latch will now remain in this state until there is a one input at D at which time the latch will toggle back the other way.

THE SET-CLEAR FLIP-FLOP

The S-C flip-flop (also known as the S-R) is a latch that has been packaged as a unit Figure 5-2. With a one at S the output Q goes to a one, and with a one at C, $\overline{Q}$ goes to a one and Q goes to zero.

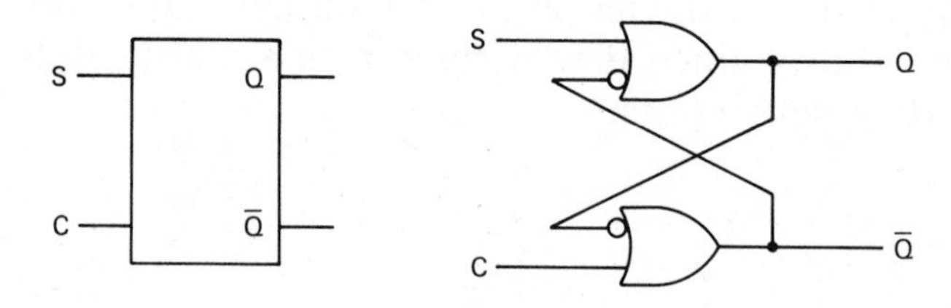

Figure 5-2

The flip-flop may also have inputs for set and clear that can override the S and C inputs. These are shown in Figure 5-3. These inputs will determine the output of the flip-flop regardless of the inputs at S and C.

CLOCKED FLIP-FLOP

The next step in the evolution of flip-flops is to add a clock input to the standard S-C flip-flop. This circuit is shown in Figure 5-4. Ths device is much more versatile than the standard S-C flip-flop. With the clock input held high

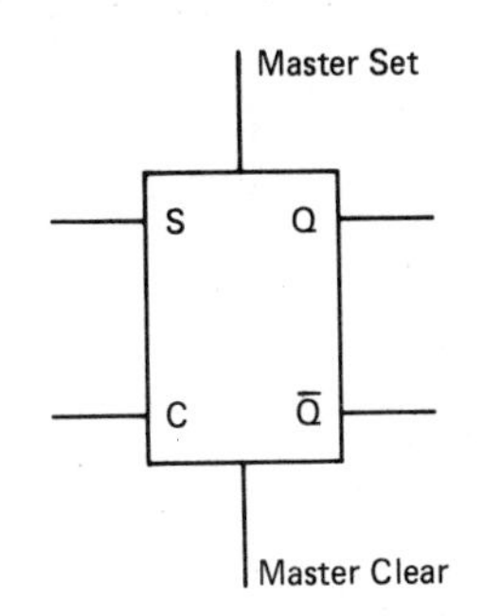

Figure 5-3

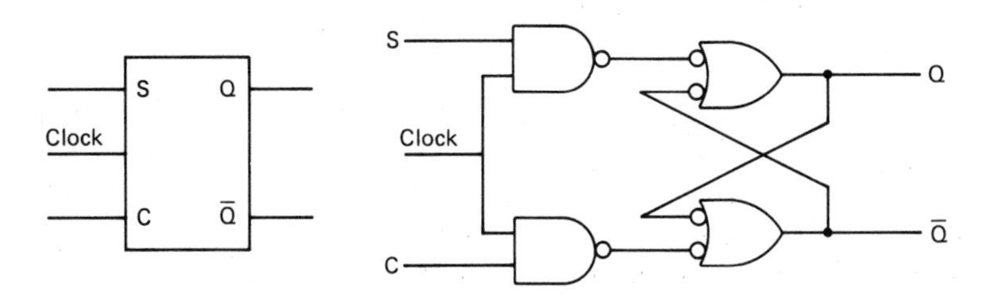

Figure 5-4

(one) the device will act just like a S-C flip-flop, but in normal operation the clock input is being pulsed. When the clock goes high (one) the output will change according to the current condition of the S and C inputs. This allows the flip-flop output to be "timed" according to some system clock.

The flip-flop of Figure 5-4 is known as a positive-edge triggered flip-flop. The output will change as soon as the clock pulse goes high or on the positive edge as shown in Figure 5-5.

This condition can be good or bad depending upon the application. Most devices are made to toggle on the negative or trailing edge of the clock pulse.

D-TYPE FLIP-FLOP

The D-type flip-flop consists of an S-C flip-flop with the S and C inputs tied together through an inverter, (see

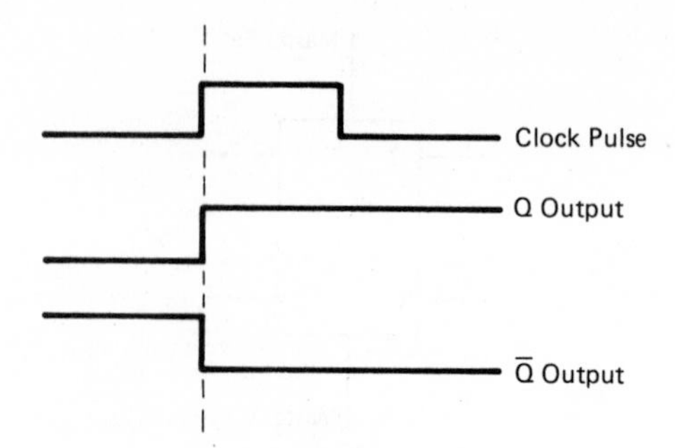

Figure 5-5

Figure 5-6). The device is used to propagate the input at D to the output at Q whenever a clock pulse is applied.

The device shown in Figure 5-6 is known as a positive-edge triggered device since transistions at the output occur on the leading edge of the clock pulse.

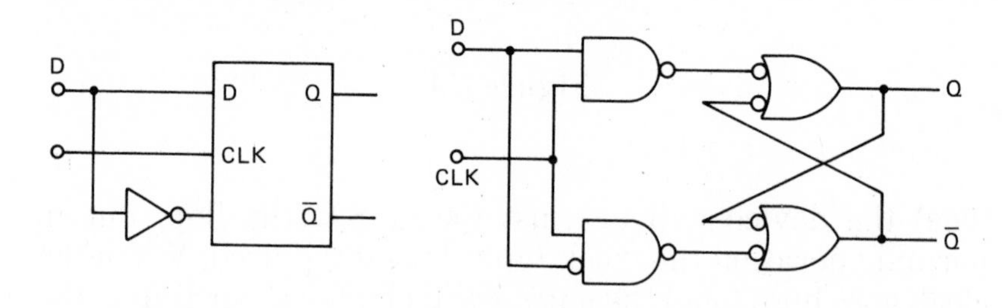

Figure 5-6

Positive-edge triggered flip-flops are available in off-the-shelf integrated circuits. An example of this is the 7477, a quadruple D-type flip-flop, all in one 14-pin integrated circuit. The circuit of the 7477 is shown in Figure 5-7.

MASTER-SLAVE J-K FLIP-FLOP

In the sequence of flip-flop complication, the master-slave J-K flip-flop is on the top of the list. A logic diagram of the device is shown in Figure 5-8.

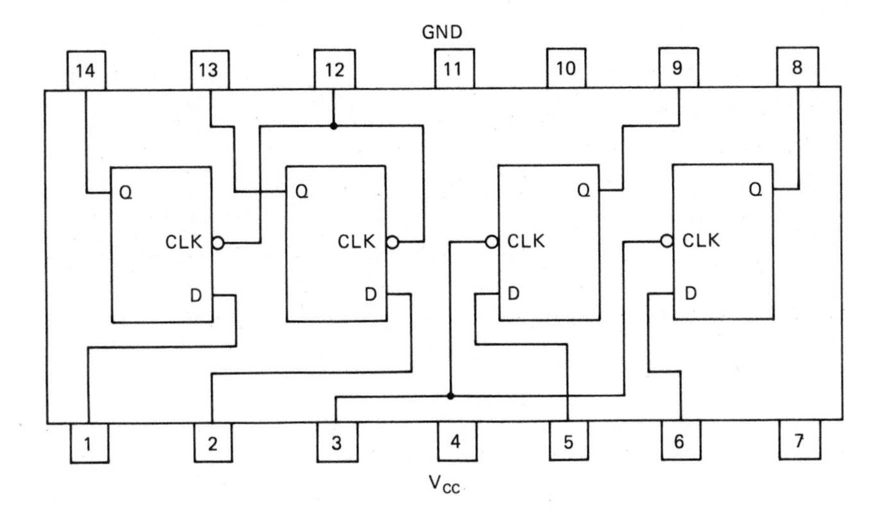

Figure 5-7

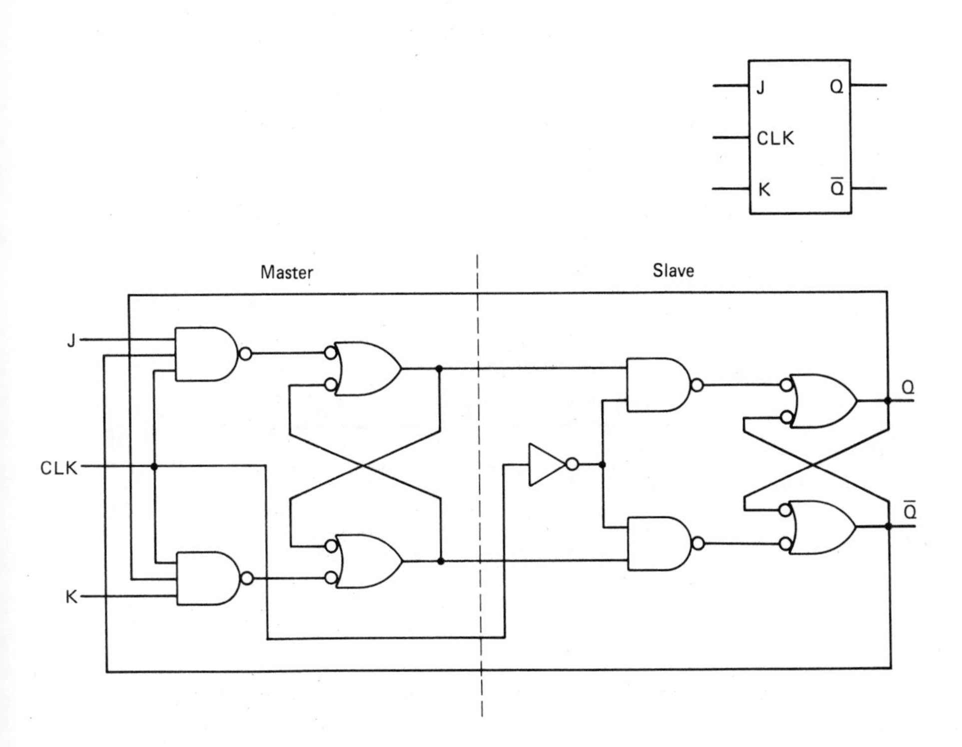

Figure 5-8

Internally the device consists of two flip-flops, a master flip-flop on the input and a slave flip-flop on the output. The master flip-flop acts like a clocked S-C flip-flop. With J high and K low the master flip-flop will toggle on the leading edge of a clock pulse. The slave flip-flop, however, will not toggle because its input depends on the master having toggled, and the clock is fed through an inverter. When the clock pulse goes back to zero, the slave flip-flop will toggle and the original input $J = 1$, $K = 0$ will propagate to the output. This makes the J-K flip-flop a negative-edge triggered device.

With the device connected as shown in Figure 5-9 A or B, the J-K is a divide by two device. Clock pulses applied with J and K held high, or applied to J and K with the clock held high, will be divided by two so that the output at Q will be one-half the input frequency. By cascading J-K flip-flops it is possible to divide the clock frequency by any power of two that is desired.

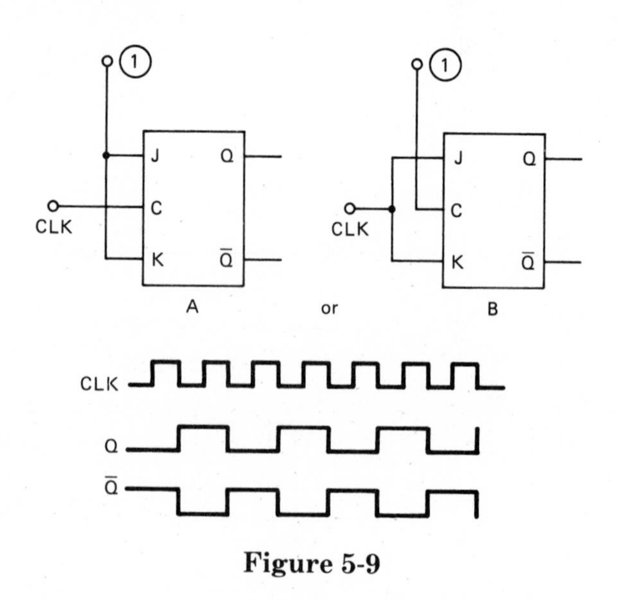

Figure 5-9

Master-slave J-K flip-flops are available in off-the-shelf integrated circuits. One example of this is the 7476. The 7476 is a dual master-slave J-K flip-flop with preset and clear inputs. The circuit of the 7476 is shown in Figure 5-10.

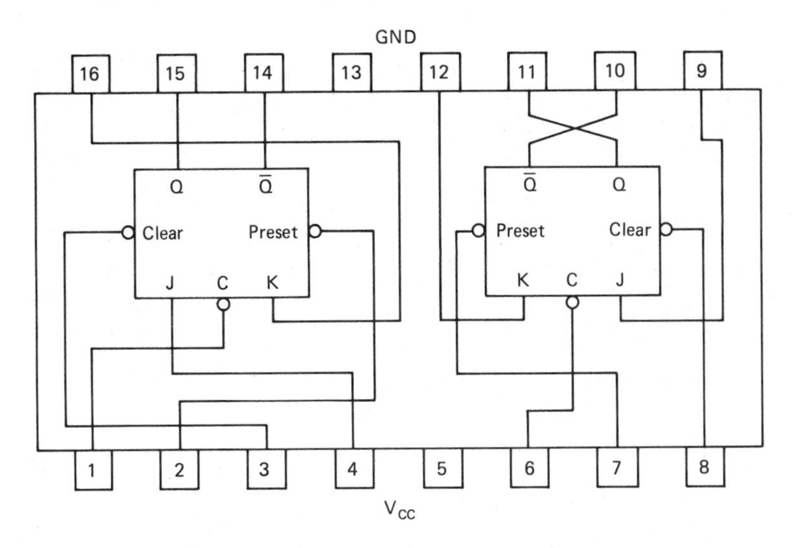

Figure 5-10: 7476 Dual J-K Flip-Flop.

BINARY COUNTERS

The circuit of Figure 5-11 shows the basic binary counter. It consists of negative-edge triggered J-K flip-flops with the Q output of each device tied to the clock input of the next device. The J and K inputs must be wired so that they are always high.

This circuit is connected as a binary "up" counter. The Q outputs of the flip-flops correspond to binary weights of 1, 2, 4, and 8 and each time a clock pulse is applied, the counter increments (counts-up) one more binary number. The waveforms given in Figure 5-11 also show:

A. Each successive flip-flop divides the previous flip-flop signal by two.
B. The output of the fourth flip-flop is one-sixteenth the frequency of the clock frequency.
C. At the same time that the Q outputs are counting up, the $\overline{Q}$ outputs are counting down.
D. It takes a total of 16 clock pulses for this four-flip-flop counter to go back to all zeros. This means that it will count 16 pulses, zero through 15.

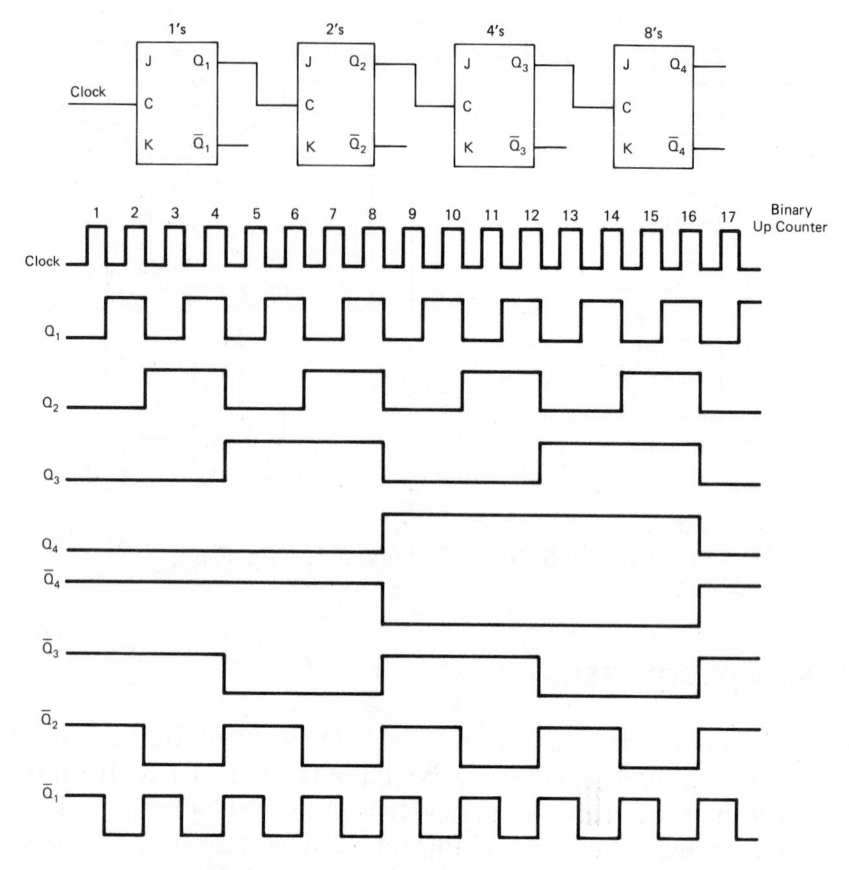

Figure 5-11: Binary Up Counter.

It's possible to construct a binary "down" counter by connecting the flip-flops as shown in Figure 5-12. Instead of feeding the Q output to the next stage, the $\overline{Q}$ output is fed to the next stage. On the negative edge of the first clock pulse all four flip-flops will toggle so that the counter is reading 15. With each clock pulse thereafter, the counter will count down by one until the counter has reached zero and then the process will start over.

From the waveforms in Figure 5-11, it is obvious that while the Q outputs are counting up the $\overline{Q}$ outputs are counting down. Similarly, the down counter of Figure 5-12

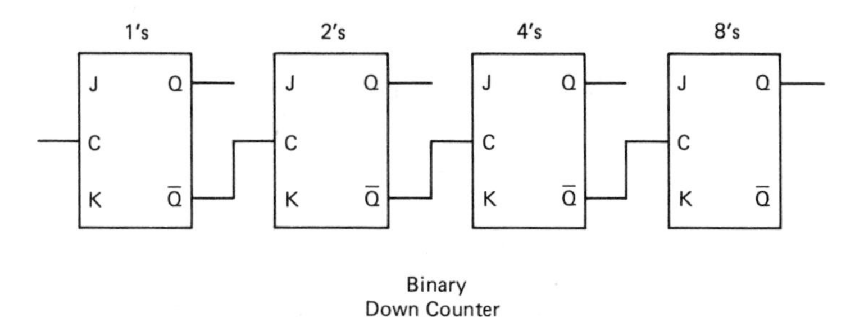

Figure 5-12: Binary Down Counter.

is also counting up and down, depending on whether one looks at the Q or $\overline{Q}$ outputs. Since it is common to consider only the Q outputs, the up and down counters are defined by the circuit connections shown in the figures. It is possible, by gating networks, to construct a binary counter that can count both up and down. One such circuit is shown in Figure 5-13. When a one is applied to the count-up line, the Q output is fed to the next flip-flop. When a one is applied to the count-down line, the $\overline{Q}$ output is fed to the next flip-flop. Changing from up to down can be done by controlling the count-up/count-down lines.

There exist several different types of binary counters some quite different from the one shown in Figure 5-11. One common type involves a shift register which is shown in a later section of this chapter. Another type of binary counter is shown in Figure 5-14. This counter uses D-type flip-flops with a feedback connection on each flip-flop.

BASE-X COUNTERS

The flip-flop lends itself readily to binary counters since its output can easily be made to be one-half of the input. In order to use flip-flops to count in different base systems, it is necessary to modify the basic binary counter. There are two basic ways of modifying the binary counter (1) the reset method and (2) the count-advance method.

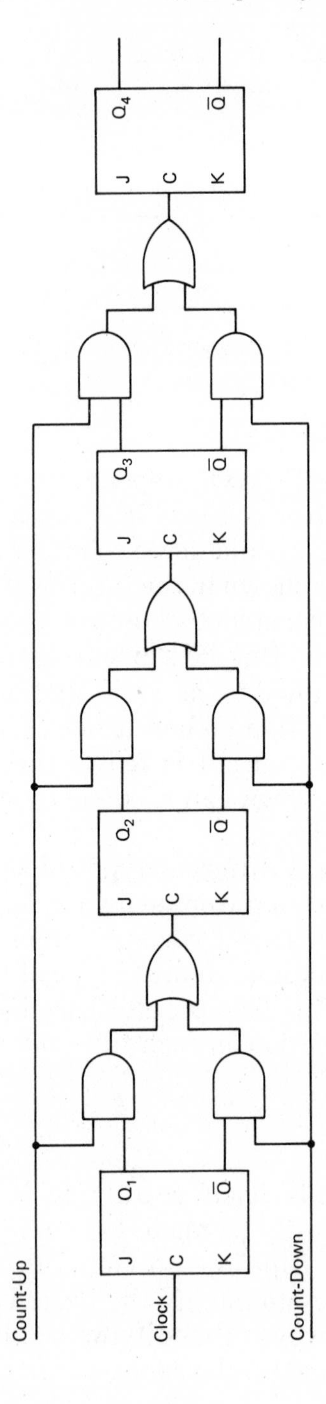

Figure 5-13: Binary Up-Down Counter.

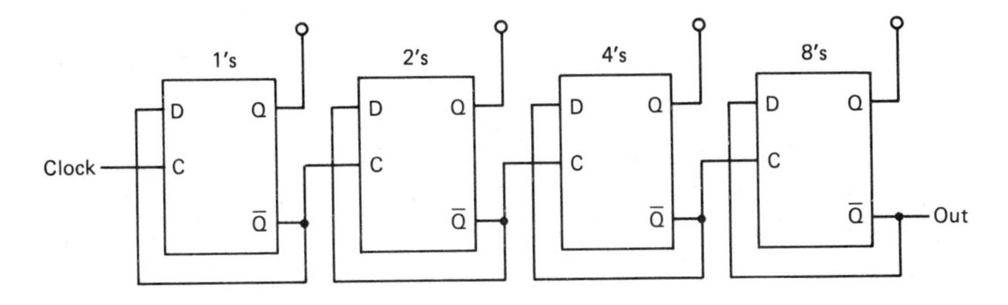

Figure 5-14: Binary Counter Using D-Type FF.

The Reset Method

The reset method involves letting the binary counter count up to the desired number and then resetting the entire counter and forcing it to start over. The flip-flops used need to have a master clear input such as described in Figure 5-3.

For an example, suppose we wanted a counter that would count in base 10 or a divide-by-ten counter. Starting with a binary counter with four stages (three stages can only count to seven) a reset is applied when ten clock pulses have been counted. This is done by connecting the Q outputs of the 2 and 8 flip-flops through an AND gate and having the output of the AND gate clear all four flip-flops. (see Figure 5-15).

This circuit gives an output pulse on the reset line for every ten clock pulses. It is therefore a count-to-ten or divide-by-ten counter. The same general concept can be used to count in any base system. For example, to count in base nine, connect the Q outputs of the 8's and 1's flip-flops to the input of the AND gate. Under this condition, all four flip-flops will be cleared after nine clock pulses.

The reset method does have some problems. The clear line will go high as soon as the Q output of the 2 flip-flops goes high. The reset line will remain high until either the 2 or 8 flip-flop resets. If, because of slight differences in devices, one of the flip-flops resets before the other, the

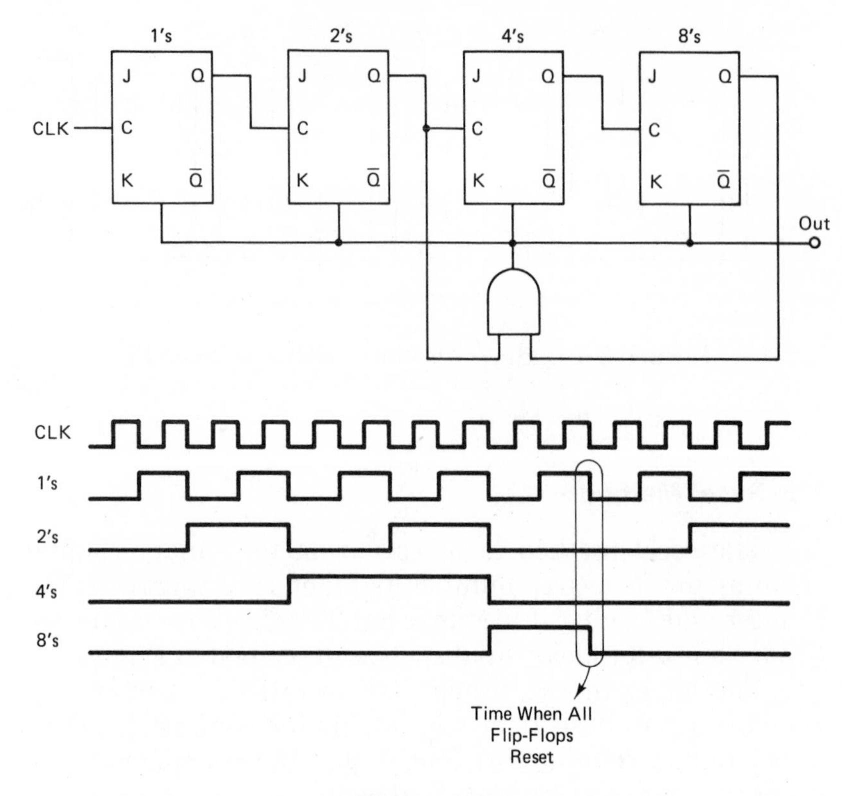

Figure 5-15

slower flip-flop would remain high. This would affect the next cycle. It is common practice to use a single-shot device (as shown in Figure 5-16) in the reset line to prevent the kind of timing glitches that can occur. The output of the single-shot must be long enough to reset all four devices and short enough to be shorter than a clock pulse.

The Count-Advance Method

Another way to force the binary counter to count in another base is to preset or count-advance the counter. Figure 5-17 shows a binary counter with a feedback line. With this connection, the 2 flip-flop will return high when the 8 flip-flop goes high. This advances the count by two. So instead of requiring 16 pulses to reset all flip-flops the circuit of Figure 5-17 requires only 14. This circuit is then a divide-by-14.

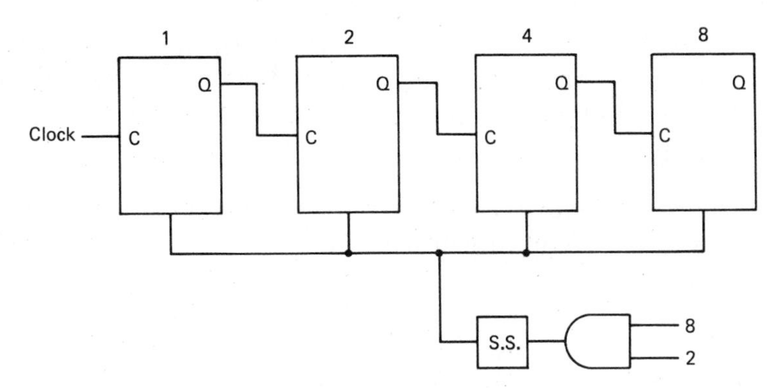

Figure 5-16

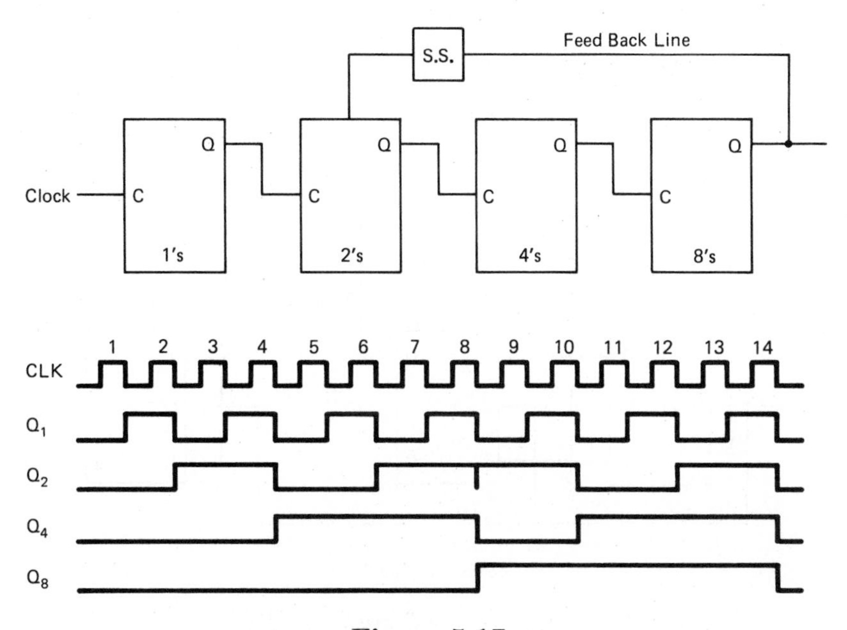

Figure 5-17

Synchronism

It is common to group counters into two categories, the synchronous counter and the asynchronous counter. The two types are defined as follows:

> *Synchronous*—In a synchronous counter, the output changes immediately after the system clock.

Asynchronous—In an asynchronous counter, the output does not change immediately after the system clock. The asynchronous counter is also known as a ripple counter.

Examples of divide-by-eight counters, one synchronous and one ripple are given in Figure 5-18.

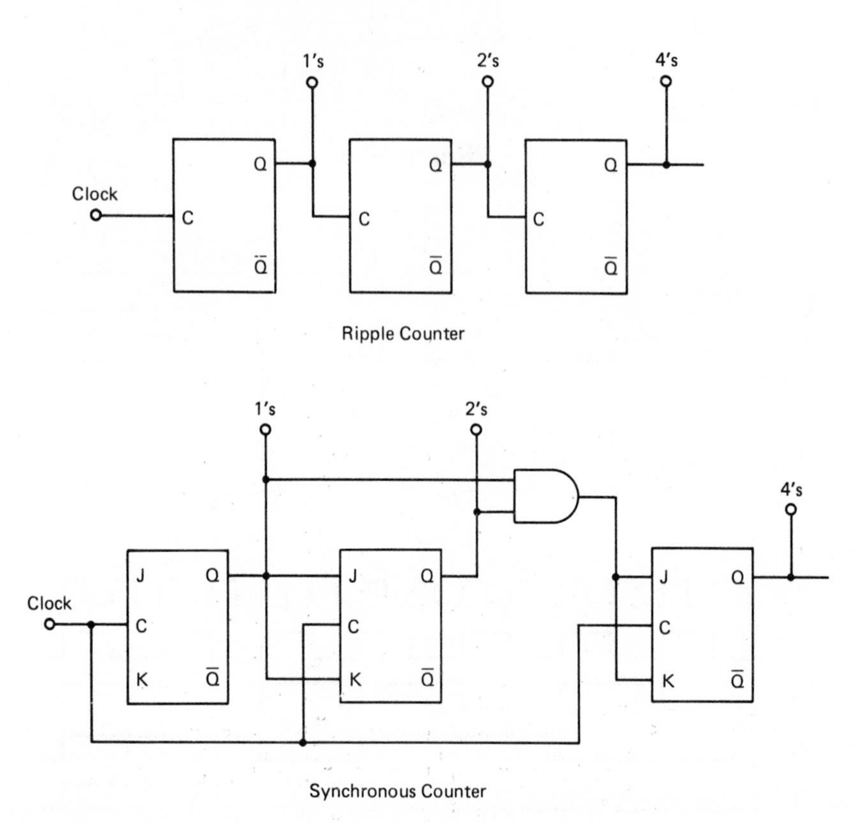

Figure 5-18: Synchronous Counter and Ripple Counter

Symmetry

Symmetry refers to the output signal of the counter. The output is symmetrical when it is high one-half of the time and low one-half of the time. The counter in Figure 5-11 has a symmetrical output while the counter shown in Figure 5-15 does not.

SHIFT REGISTERS

The basic shift register is shown in Figure 5-19. It consists of J-K flip-flops with the output of the flip-flop on the left tied to the input of the flip-flop on the right. The clock inputs are all tied together so that the clock pulse is fed to all flip-flops at the same time.

To understand how the shift register works, imagine that the Q outputs of all flip-flops are initially at zero and the data input contains a one. On the trailing edge of the first clock pulse F_1 will toggle. The register will now contain 1, 0, 0, 0. Now let the data input go to zero and apply a second clock pulse. On the trailing edge of the second clock pulse, F_1 and F_2 will toggle and the register will contain 0, 1, 0, 0. On the trailing edge of each additional clock pulse, the one will "shift" to the next flip-flop on the right until it is shifted completely out of the register.

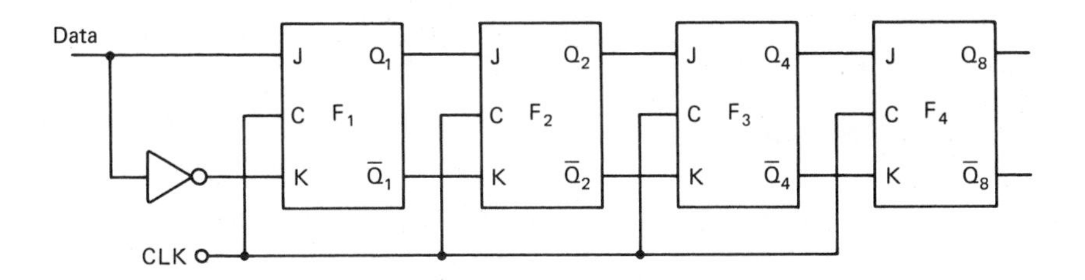

CLK	F_1	F_2	F_3	F_4
1	1	0	0	0
2	0	1	0	0
3	0	0	1	0
4	0	0	0	1

Figure 5-19: Shift Register Using JK FF Negative Edge Trigger.

Shift Register Applications

When all the digits of a binary number are shifted one place to the right, it is the same as dividing the number by two. Consider the following example:

0 1 1 0 1 0 equals 26_{10}

↓ ↓ ↓ ↓ ↓

0 0 1 1 0 1 equals 13_{10}

Each digit shifted
right one place

When each digit of the binary number 26 is shifted one place to the right, the resulting number is one-half 26 or 13. If the binary number is shifted two places to the right, the number will be divided by four, three shifts to the right is division by eight, etc.

Shifting the binary number to the left one place is the same as multiplying the number by two. Shifting two places to the left is multiplication by four, etc. Division and multiplication by powers of two can be accomplished by shifting the number to the right or left in a register. Shift-right and shift-left registers are shown in Figure 5-20.

The Ring Counter

A shift register can be converted to a ring counter by connecting the output of the last flip-flop to the input of the first flip-flop (see Figure 5-21). As the name implies, the ring counter will shift bits continuously around the ring.

One practical application of a ring counter is shown in Figure 5-22. This circuit performs a divide-by-ten function using a ring counter. If one of the flip-flops forming the ring counter contains a 1, this 1 will be shifted to flip-flop "E" every fifth clock pulse. The Q output of flip-flop E divides the clock frequency by five. This output if fed to flip-flop F which further divides the clock frequency. The output of F is one-tenth the clock frequency and the output is also symmetrical.

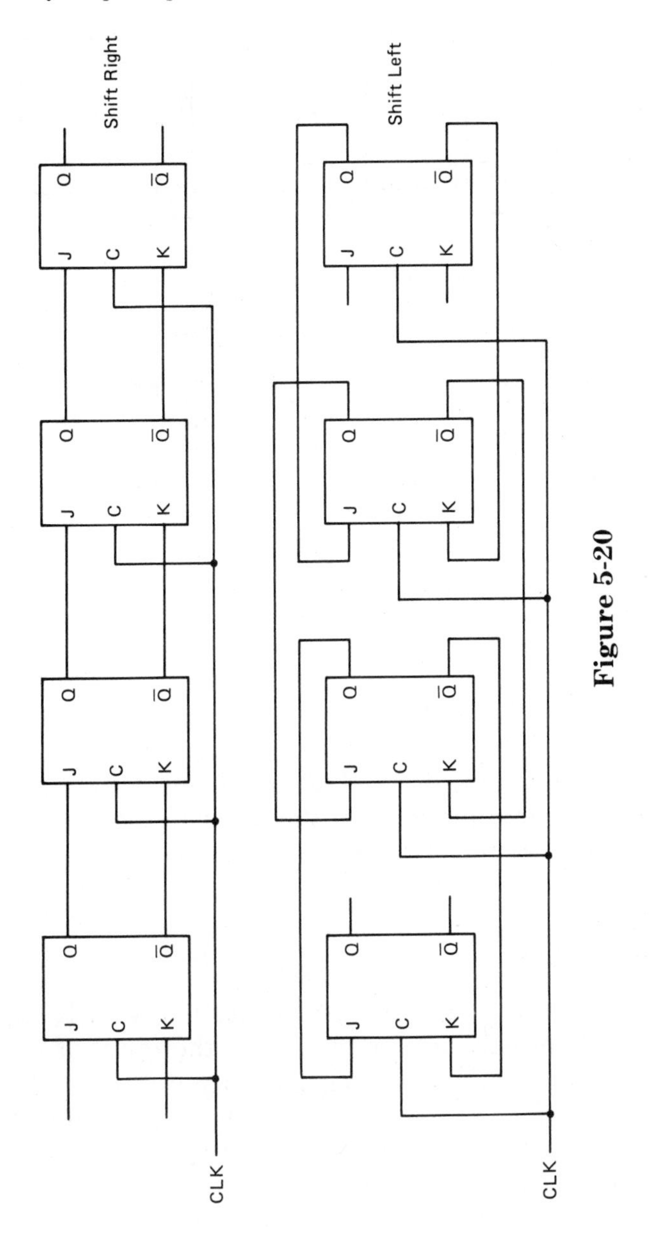

Figure 5-20

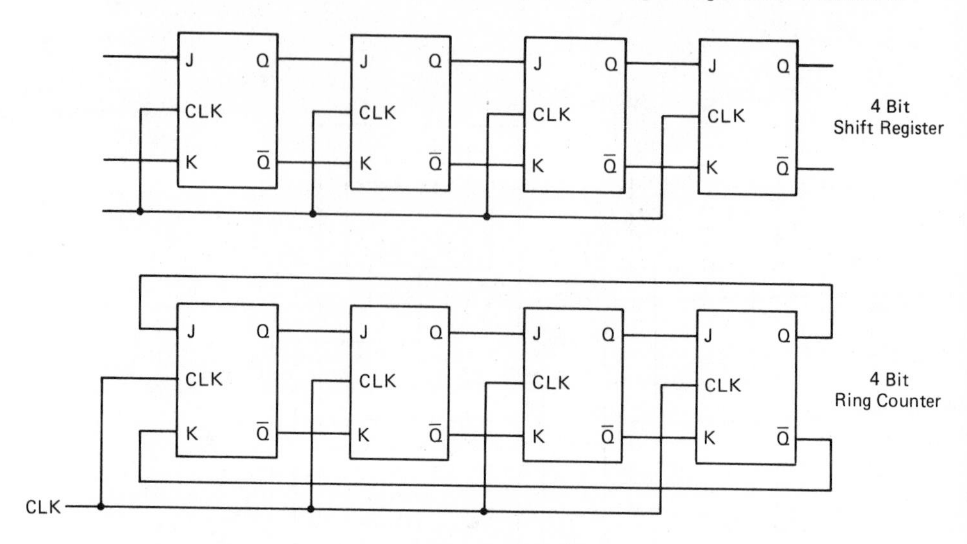

Figure 5-21

COUNTER APPLICATIONS

Flip-flops show up in all kinds of electronic equipment. Voltmeters, counters, clocks and other devices use the flip-flop to make them work. The following are some examples of the uses of flip-flops.

90° Phase Shift

The circuit of Figure 5-23 contains two J-K flip-flops the outputs of each are connected to the input of the other with one important variation. The output of F_2 is connected to the input of F_1 with a twist ($\overline{Q}_2$ connected to J and Q_2 to K). The outputs of this circuit at Q_1 and Q_2 are one-fourth the input clock frequency. The more important aspect of the outputs is that they are exactly 90 degrees out of phase. This condition is useful in certain clocking applications and in certain single-sideband applications.

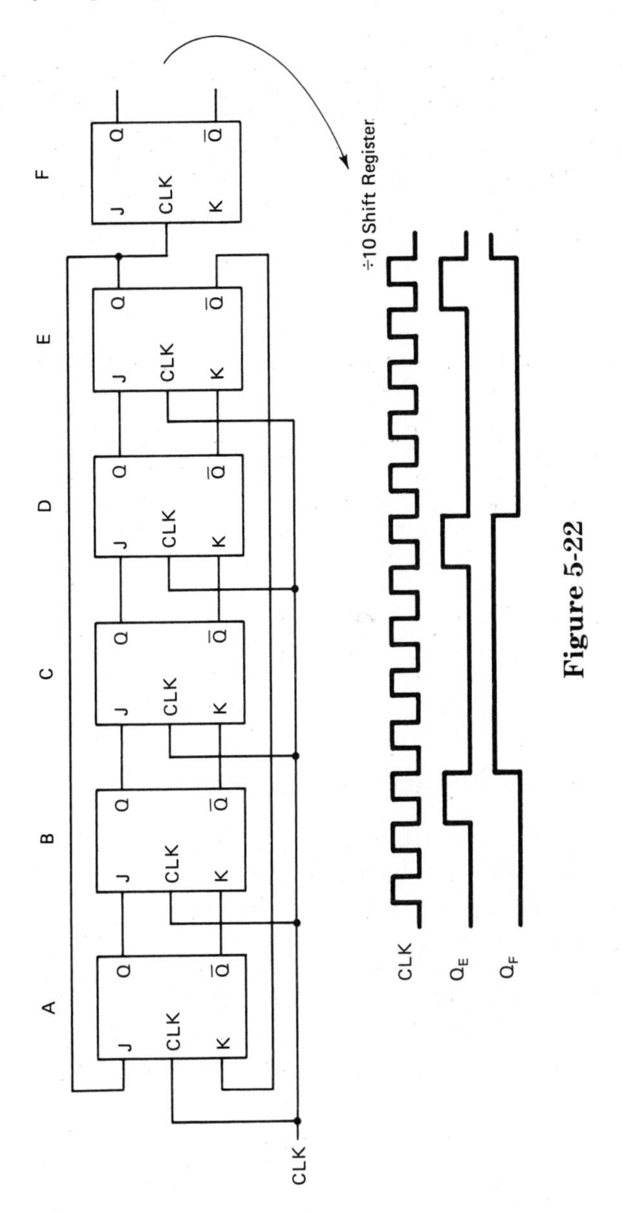

Figure 5-22

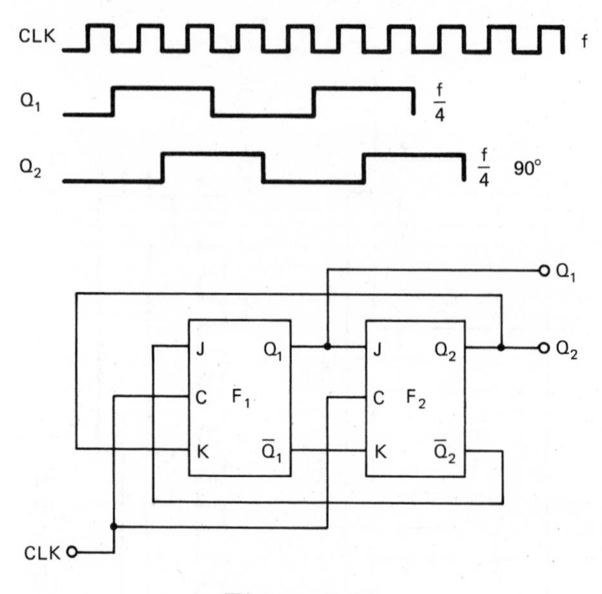

Figure 5-23

The Frequency Counter

Frequency counters use flip-flops to both count the incoming pulses and to generate precise pulse lengths to gate the counter. Figure 5-24 shows the basic circuit used in a frequency counter.

A. The count function is done by four stages of divide-by-ten. The standard 7490 IC chip serves this purpose well. This four-stage counter can count input pulses from zero to 9999.
B. The count is displayed by the use of seven segment LED displays. To decode the binary output of the 7490 counters into the seven segment LED inputs, 7447 IC chips are used. These chips were designed especially for this purpose.
C. The count and display section will continue to function as long as the input frequency can pass through the AND gate. This is determined by placing a pulse on the other input of the AND gate. This pulse width

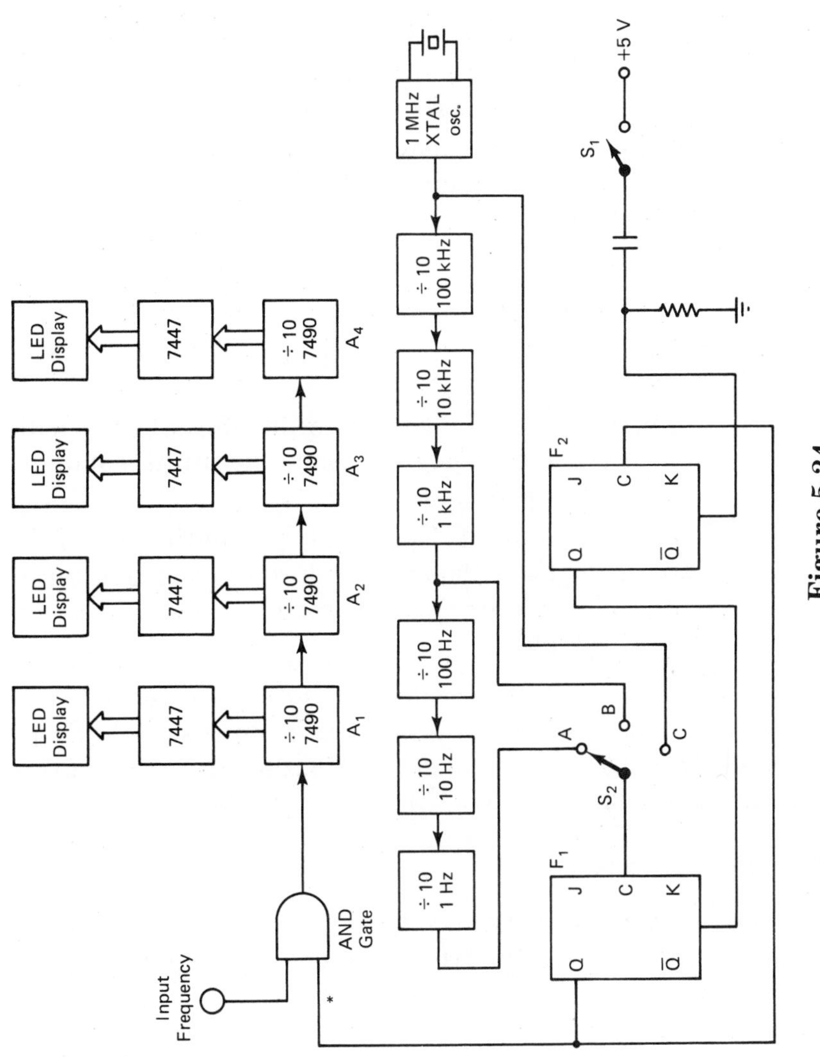
LED Display
LED Display
LED Display
LED Display
7447
7447
7447
7447
÷ 10 7490
÷ 10 7490
÷ 10 7490
÷ 10 7490
A_1
A_2
A_3
A_4
Input Frequency
AND Gate
*
1 MHz XTAL osc.
÷ 10 100 kHz
÷ 10 10 kHz
÷ 10 1 kHz
÷ 10 100 Hz
÷ 10 10 Hz
÷ 10 1 Hz
A
B
C
S_2
S_1
+5 V
F_1
J
C
K
Q
$\bar{Q}$
F_2
J
C
K
Q
$\bar{Q}$

Figure 5-24

determines the range of frequencies counted. For example:

1. A one second pulse means that the pulses are counted in hertz (Hz) (cycles per second).
2. A one millisecond pulse means that the pulses are counted in hertz (KHz) (kilocycles per second).
3. A one microsecond pulse means that the pulses are counted in MHz (megacycles per second), etc.

D. The precise pulse length is supplied to the AND gate by a crystal oscillator and a series of divide-by-ten flip-flops.

E. The switch S_2 can select one of three frequencies:

1. One MHz
2. One kHz
3. One Hz.

Since F_1 is a divide-by-two, the output at Q will be a pulse either one second, one millisecond, or one microsecond in length.

F. The pulses out of F_1 would be continuous if it weren't for F_2. F_2 holds F_1 in a reset condition until switch S_1 is closed. This resets F_2 and allows F_1 to put out one pulse.

The frequency counter counts for either one second, one millisecond, or one microsecond each time switch S_1 is pressed. In most counters, the function of S_1 is automatic, but the basic principles are the same.

The Digital Clock

The digital clock is also a big user of flip-flops. It is basically the same as the frequency counter except that it must count in base 60 and base 12 to properly count seconds, minutes, and hours.

The block diagram of a digital clock is shown in Figure 5-25. The counters must count in base 60 and 12. A count-advance type of base-60 counter is shown in Figure 5-26.

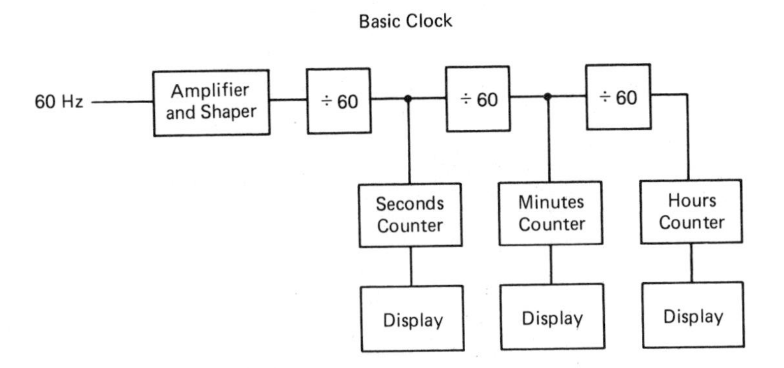

Figure 5-25: Basic Clock.

A. The reference input in this circuit is the 60-Hz frequency of normal house current. Some clocks use an internal crystal oscillator as the reference.

B. The 60-Hz sinewave must be half-wave rectified and shaped into square pulses with +5.0 volts amplitude.

C. The first divide-by-sixty provides one pulse every second. Its output is then fed to a counter and displayed if the clock has a seconds readout.

D. The second divide-by-sixty provides one pulse every minute. Its output feeds the minute counter and display.

E. The third divide-by-sixty provided one pulse every hour and reads the hour counter and display.

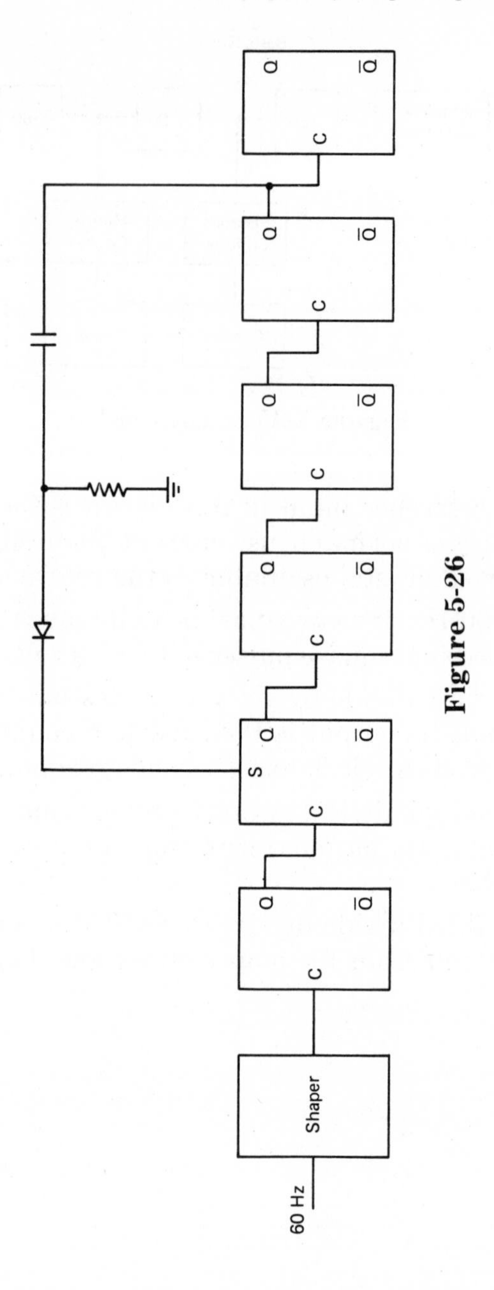

Figure 5-26

6

Understanding Microprocessors

INTRODUCTION

Modern microprocessors evolved over time from circuits such as Arithmetic Logic Units (ALUs), registers and shift registers, timing and control circuits, etc. In the simplest form, a microprocessor can be viewed as a black box with lines connected to it as shown in Figure 6-1. These lines are common to all microprocessors. The data lines allow the input of data from external devices and the output of data from the microprocessor. The address lines control which device is being communicated with. The control lines perform timing, reset and read/write functions.

BUILDING A MICROPROCESSOR

To understand the microprocessor functions, we will start with the circuit shown in Figure 6-2. This circuit is almost the same as the one shown in Chapter 4, Figure 4-4, except that instead of a third register to accept the output of the ALU, the output is fed back to the accumulator register. The accumulator register becomes a very important part of microprocessors. It is the primary register in most devices and almost all operations involve the accumulator register.

The circuit of Figure 6-2 is capable of taking two eight-bit binary numbers from the registers, performing various arithmetic operations on them, and feeding the results back

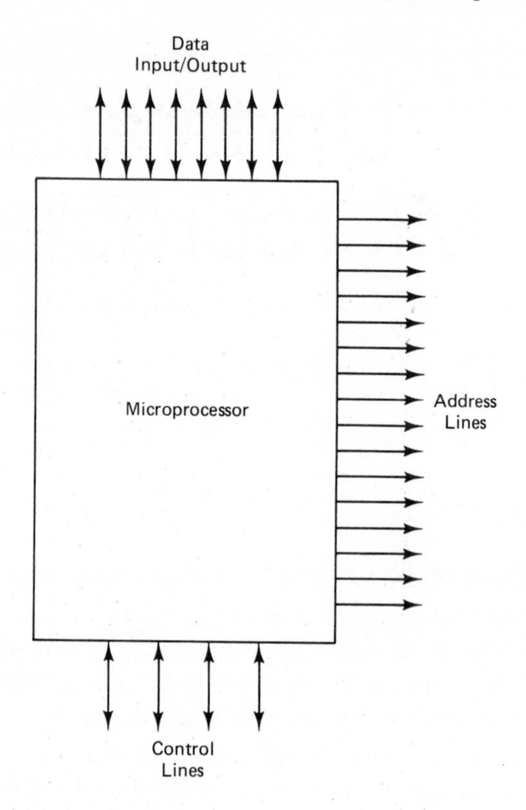

Figure 6-1: A Microprocessor as a Black Box.

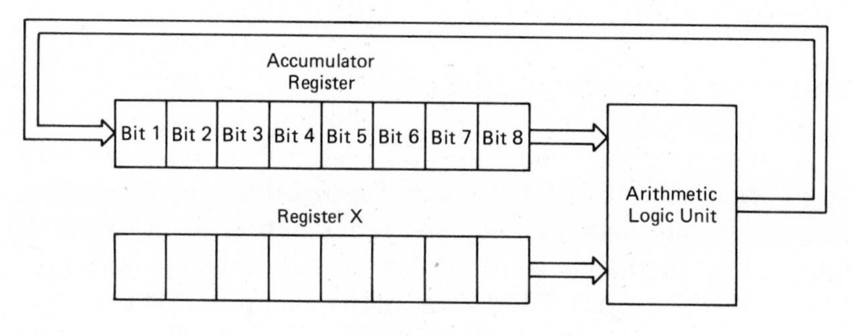

Figure 6-2

into the accumulator register. The arithmetic operations consist of addition, subtraction and various other operations. The ALU is the heart of the microprocessor and ICs are available that are just ALUs.

The Arithmetic Logic Unit (ALU)

A typical ALU chip is the 74181 IC. This device has inputs for two four-bit binary numbers and can perform various arithmetic operations. The arithmetic operation to be performed is controlled by a four-bit binary number inputed to the chip on four separate lines. These four-bit numbers form a simple instruction code set similar to larger instruction code sets for actual microprocessors. The pin connection and instruction codes for the 74181 are shown in Figure 6-3. In order to add the two numbers, the instruction code on the S inputs must be 1001 and in order to subtract, the instruction code must be 0110.

The ALU performs the operation defined by the instruction code because this input enables the appropriate gate inside the ALU. By so directing the two data inputs, the various arithmetic operations can be performed. The internal gating of the 74181 is shown in Figure 6-4.

A microprocessor is capable of performing as many operations as the instruction set will allow. It performs these instructions in a serial form, one after the other. The sequence of instructions is stored in memory which is usually external to the microprocesor. They are brought, one at a time, into the instruction register. This is internal to the microprocessor.

The Program Memory and Program Counter

Figure 6-5 shows the next step in the building of a microprocessor. There is a memory section which contains the sequence of instructions for the arithmetic logic unit. There is an instruction register in which the instruction codes from memory are loaded one at a time. The instruction in the instruction register is the one that the ALU is performing. There is another register called the program counter. The program counter increments by one each time an instruction has been completed. This increment by one is used to "pick" the next sequential instruction in program memory and feed it to the instruction register. The program counter will continue to pick sequential memory locations until all of the instructions have been executed.

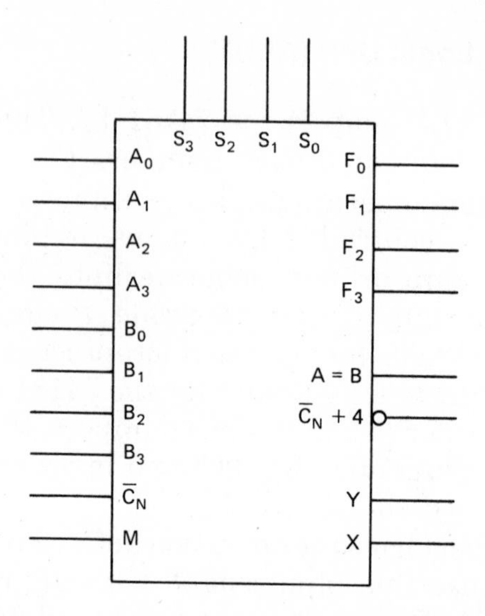

Operation Code				Arithmetic Operations	
S_3	S_2	S_1	S_0	$\overline{C}_N = 1$ (No Carry)	$\overline{C}_N = 0$ (With Carry)
0	0	0	0	F = A	F = A Plus 1
0	0	0	1	F = A + B	F = (A + B) Plus 1
0	0	1	0	$F = A + \overline{B}$	$F = (A + \overline{B})$ Plus 1
0	0	1	1	F = Minus 1	F = Zero
0	1	0	0	F = A Plus $A\overline{B}$	F = A Plus $A\overline{B}$ Plus 1
0	1	0	1	F = (A + B) Plus $A\overline{B}$	F = (A + B) Plus $A\overline{B}$ Plus 1
0	1	1	0	F = A Minus B Minus 1	F = A Minus B
0	1	1	1	F = $A\overline{B}$ Minus 1	F = $A\overline{B}$
1	0	0	0	F = A Plus AB	F = A Plus AB Plus 1
1	0	0	1	F = A Plus B	F = A Plus B Plus 1
1	0	1	0	F = $(A + \overline{B})$ Plus AB	F = $(A + \overline{B})$ Plus AB Plus 1
1	0	1	1	F = AB Minus 1	F = AB
1	1	0	0	F = A Plus A*	F = A Plus A Plus 1
1	1	0	1	F = (A + B) Plus A	F = (A + B) Plus A Plus 1
1	1	1	0	F = $(A + \overline{B})$ Plus A	F = $(A + \overline{B})$ Plus A Plus 1
1	1	1	1	F = A Minus 1	F = A

Figure 6-3

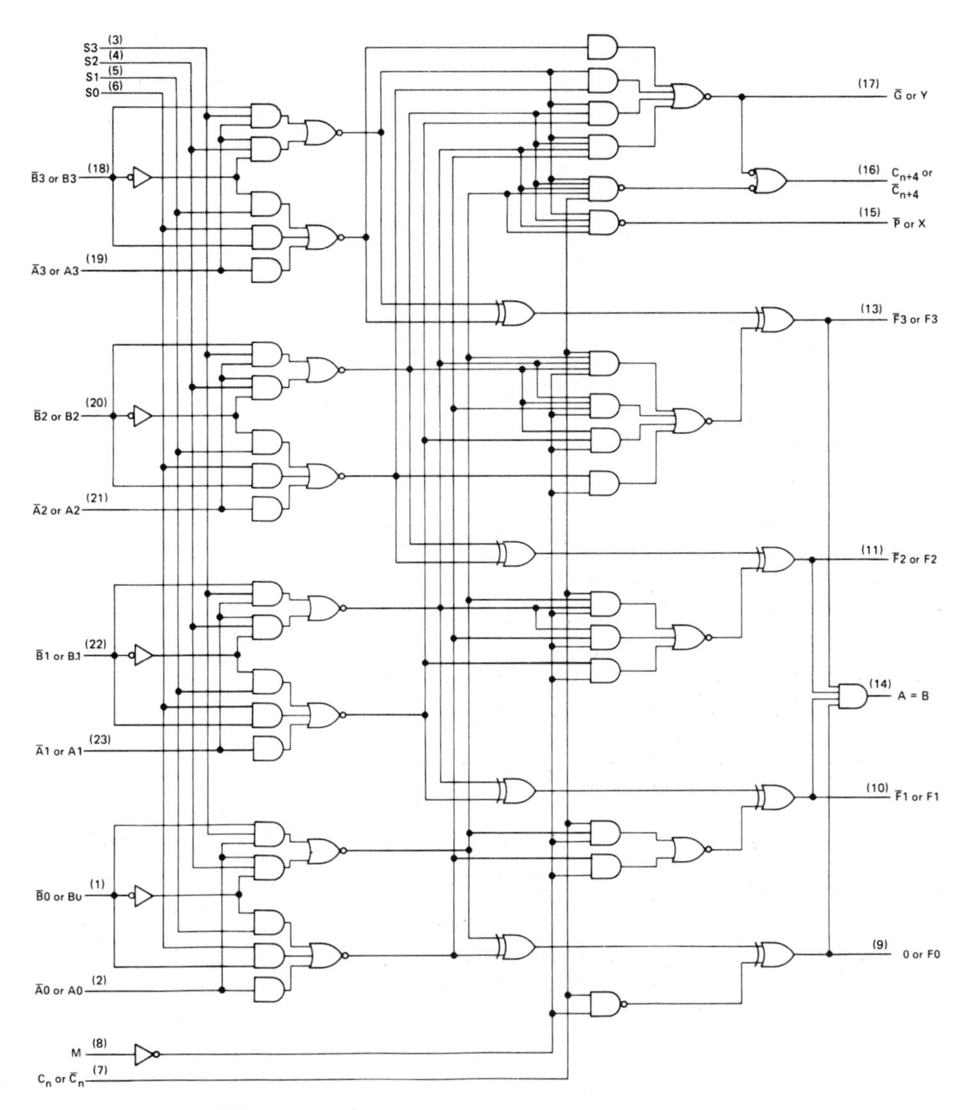

Figure 6-4: Internal Gating of the 74181.

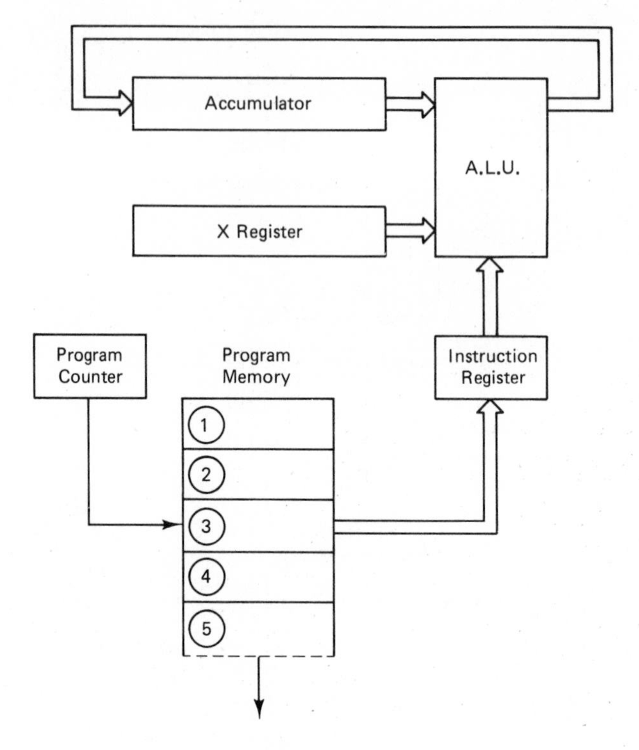

Figure 6-5

Timing and Control

The one item missing from Figure 6-5 that is necessary to have a complete microprocessor is the timing and control function. To shift instructions into the instruction register, to increment the program counter, and to move data around in general, it is necessary to have a master clock and special circuits to handle the timing.

Figure 6-6 shows the next step in the building of a microprocessor. It contains a timing and control block which is driven by a crystal oscillator (the clock). For an example of the control and timing function, consider the example of adding the contents of the accumulator to the contents of the X register. The sequence of control and timing goes like this:

1. The add instruction is placed into the instruction register from the memory address called out by the

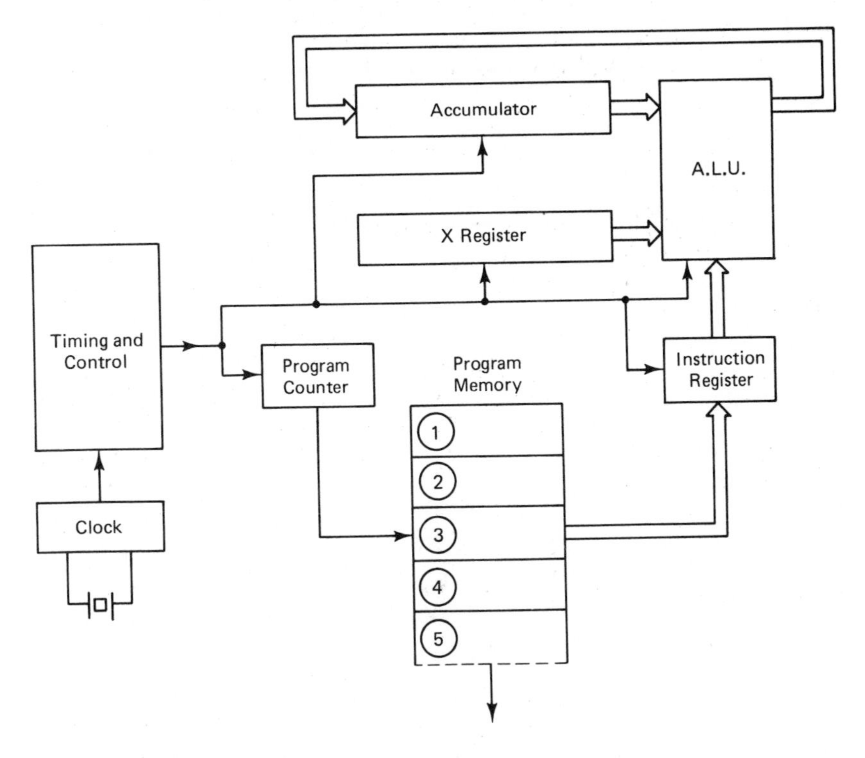

Figure 6-6

program counter.

2. The time interval pulse for addition is generated by the timing and control section. This pulse is timed from the basic clock as shown in Figure 6-7.
3. During the time interval pulse, clock pulses are applied to the two registers in order to shift the data through the ALU and back into the accumulator.
4. Because the ALU was programmed to add, the number now in the accumulator is the sum of the two registers. The X register now contains zero.
5. At the end of the time interval pulse is another pulse coming from the timing and control section. This pulse increments the program counter. Once incremented, the program counter selects the next sequential address in program memory. With proper timing from the timing and control section, the next instruction will be executed.

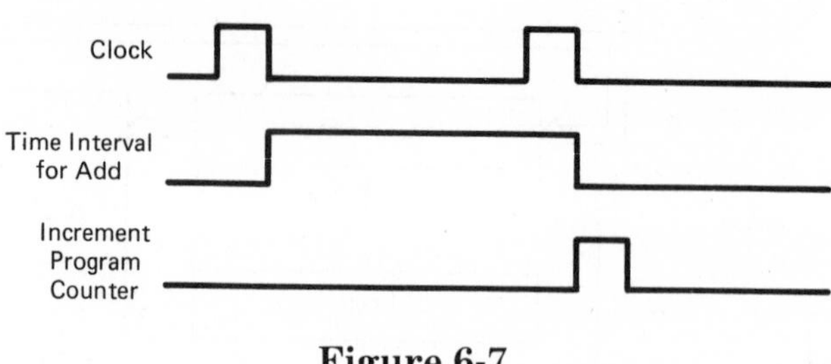

Figure 6-7

INSTRUCTION WORD FORMATS

Up till now, we have assumed that the instructions loaded into the instruction register are single eight-bit binary words. This is not always the case in real microprocessors. The simplest instruction is the single-word format. This consists of an eight-bit "op" code as shown in Figure 6-8(A). The op code is the binary instruction that tells the ALU what to do (add, subtract, etc.).

In a two-word instruction as shown in Figure 6-8(B), the two eight-bit words are stored in sequential addresses. The op code is first and again is the binary instruction that tells the ALU what to do. The second eight-bit word is either data to be used by the ALU or it is a memory location. The memory location called out contains the data that is to be processed per the instructions of the op code.

The three-word instruction shown in Figure 6-8(C) is just a slight variation from the first two. The first eight-bit word is still the op code. The second two eight-bit words form a 16-bit address of the memory location in which the data is stored. The first eight-bit word contains the least significant half of the address and the second eight-bit word contains the most significant half. In general, there are several different types of addressing modes as they concern program counters and memory. The major modes are shown in the following table. In each case, the program counter points to the op code.

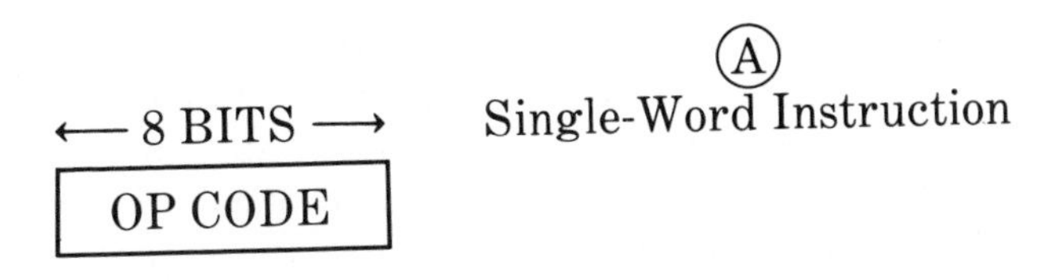

Ⓑ
Two-Word Instruction

OP CODE
ADDRESS OR DATA

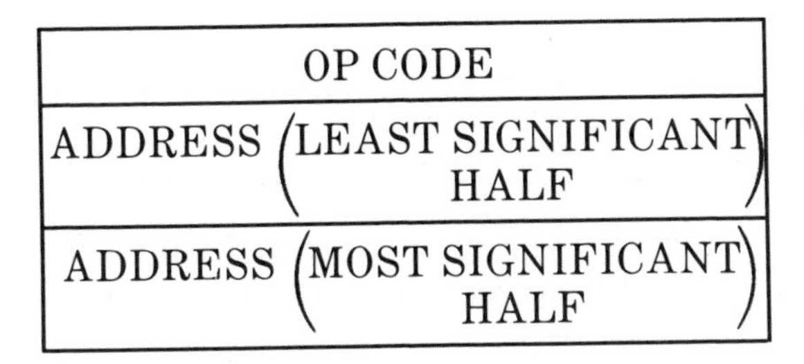

Ⓒ
Three-Word Instruction

Figure 6-8

ADDRESSING MODE	PROCESSING
Immediate	Current value of the program counter indicates the op code. The digital information represented by PC + 1 is the data the op code is to perform its operation on.
Direct	Current value of the program counter indicates the op code. Increment the program counter and then move the data from the location specified by program counter plus 1 to the address register.

Extended	Current value of the program counter indicates the op code. Increment the program counter and transfer the data from the location specified by program counter plus 1 to the address register. Increment the program counter again and transfer the data from the location specified by program counter plus 2 to the address register.
Relative	Current value of the program counter indicates the op code. Increment the program counter by 1 and add the contents of the location specified by program counter plus 1 to the value of the program counter after it is incremented again.
Indexed	Current value of the program counter indicates the op code. Increment the program counter by 1 and add the contents of the location specified by program counter plus 1 to the index register.

EXAMPLE OF A "WORKED OUT" PROGRAM

In order to see how a microprocessor actually performs a function, it is necessary to step through a sample program. The sample program for this example will be the addition of two numbers. The instructions used in the example will be actual instruction codes for the 8080 microprocessor. The instructions are:

1. *Load Accumulator*, (00111010)—This is a three-byte instruction that takes the contents of the address specified by the second and third bytes and loads it into the accumulator.
2. *Add Immediate*, (11000110)—This is a two-byte instruction that takes the contents of the second byte of the instruction and adds it to the content of the accumulator. The sum is stored in the accumulator replacing the previous number.

3. *Halt,* (01110110)—This is a single-byte instruction that stops all processing.

These three instructions incorporate one each of the type of instruction words shown in Figure 6-8.

The program for the addition of the two numbers 55 and 38 is shown in Figure 6-9. The first memory location contains the instruction load accumulator. The next two memory locations contain the memory address of the number to be loaded into the accumulator. This address is 1010 and the number 55 is located there. The third memory location contains the instruction "add immediate." The fourth memory contains the binary number to be added, 38, since "add immediate" is a two-byte instruction. The fifth memory location contains the instruction "halt." This instruction stops all processing. At this point in time the binary number 93 will appear in the accumulator and the program will be complete.

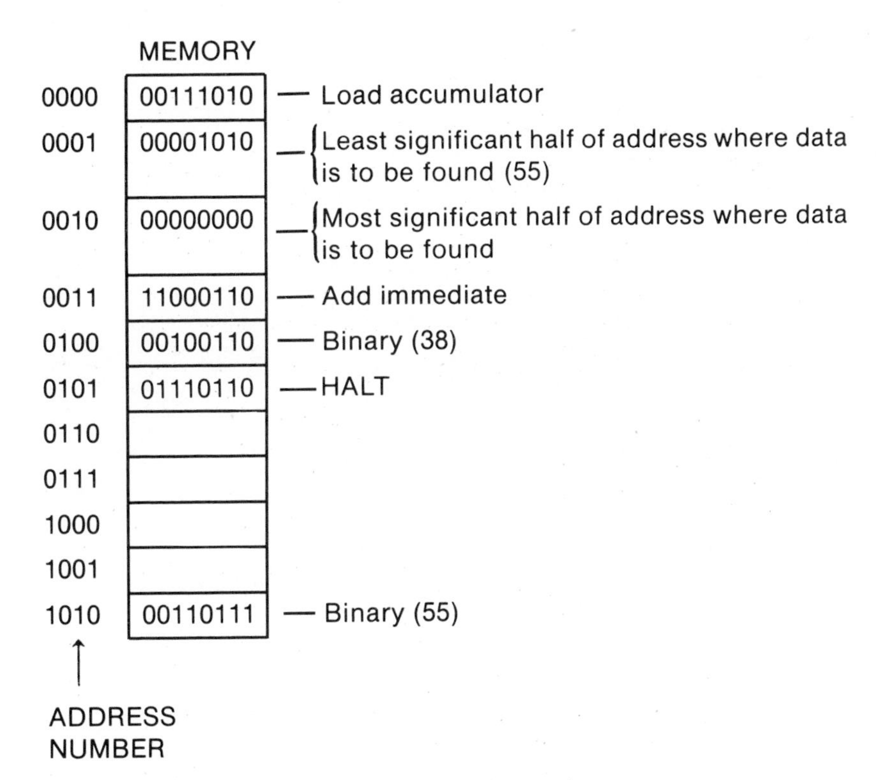

Figure 6-9: Sample Program for the Addition of 55 and 38.

Another "Worked Out" Program

The previous example added two numbers, left the sum in the accumulator, and then halted the program. There is also an instruction in the 8080 instruction set called Subtract Immediate. This is the Eight-Bit Binary Number 11010110. Figure 6-10 shows a program which adds the number 38 and 55 as in the previous example and then subtracts 12 from the result. At the end of this program, the number 81 will be stored in the accumulator.

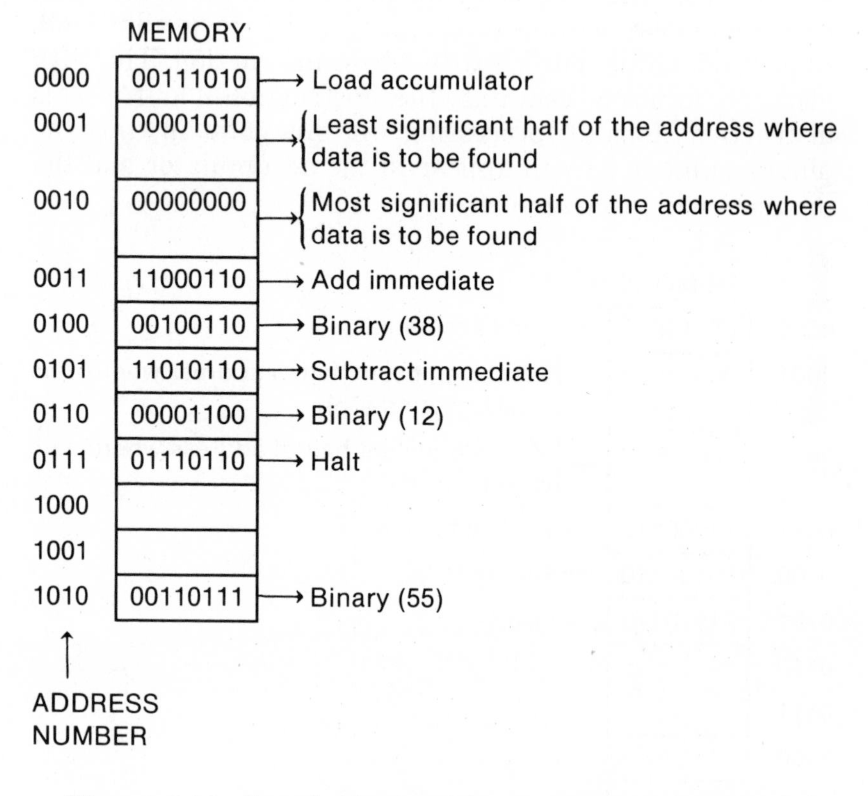

Figure 6-10: Sample Program for the Addition and Subtraction of Numbers.

THE 8080 MICROPROCESSOR

The 8080 microprocessor was one of the first eight-bit microprocessors available. It contains all of the building clocks discussed earlier in this chapter plus many more. In particular, there are many more registers than existed in previous examples (see Figure 6-11). The instruction set for the 8080 microprocessor is shown in Figure 6-12.

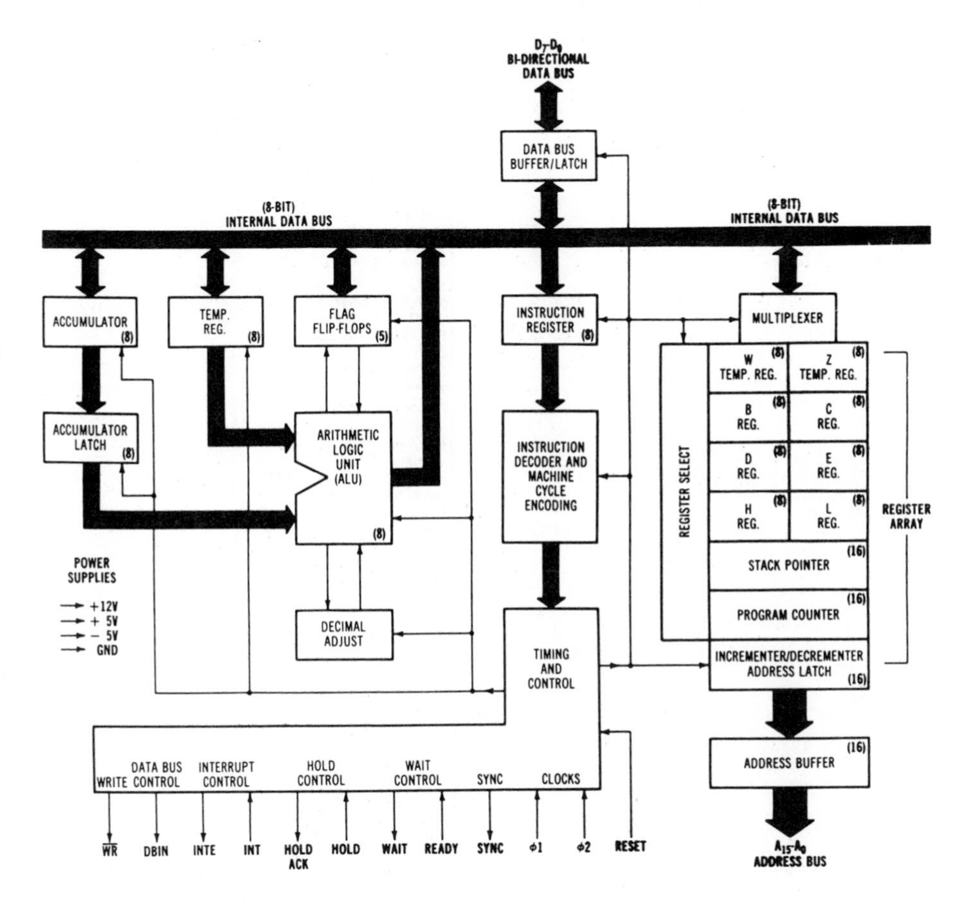

Figure 6-11: Block Diagram of the 8080A Microprocessor. Courtesy Intel Corporation.

	Instruction Code [1]								Operations	Clock Cycles
Mnemonic	D_7	D_6	D_5	D_4	D_3	D_2	D_1	D_0	Description	[2]
MOVE, LOAD, AND STORE										
MOVr1,r2	0	1	D	D	D	S	S	S	Move register to register	5
MOV M,r	0	1	1	1	0	S	S	S	Move register to memory	7
MOV r,M	0	1	D	D	D	1	1	0	Move memory to register	7
MVI r	0	0	D	D	D	1	1	0	Move immediate register	7
MVI M	0	0	1	1	0	1	1	0	Move immediate memory	10
LXI B	0	0	0	0	0	0	0	1	Load immediate register Pair B & C	10
LXI D	0	0	0	1	0	0	0	1	Load immediate register Pair D & E	10
LXI H	0	0	1	0	0	0	0	1	Load immediate register Pair H & L	10
STAX B	0	0	0	0	0	0	1	0	Store A indirect	7
STAX D	0	0	0	1	0	0	1	0	Store A indirect	7
LDAX B	0	0	0	0	1	0	1	0	Load A indirect	7
LDAX D	0	0	0	1	1	0	1	0	Load A indirect	7
STA	0	0	1	1	0	0	1	0	Store A direct	13
LDA	0	0	1	1	1	0	1	0	Load A direct	13
SHLD	0	0	1	0	0	0	1	0	Store H & L direct	16
LHLD	0	0	1	0	1	0	1	0	Load H & L direct	16
XCHG	1	1	1	0	1	0	1	1	Exchange D & E, H & L Registers	4
STACK OPS										
PUSH B	1	1	0	0	0	1	0	1	Push register Pair B & C on stack	11
PUSH D	1	1	0	1	0	1	0	1	Push register Pair D & E on stack	11
PUSH H	1	1	1	0	0	1	0	1	Push register Pair H & L on stack	11
PUSH PSW	1	1	1	1	0	1	0	1	Push A and Flags on stack	11
POP B	1	1	0	0	0	0	0	1	Pop register Pair B & C off stack	10
POP D	1	1	0	1	0	0	0	1	Pop register Pair D & E off stack	10
POP H	1	1	1	0	0	0	0	1	Pop register Pair H & L off stack	10
POP PSW	1	1	1	1	0	0	0	1	Pop A and Flags off stack	10
XTHL	1	1	1	0	0	0	1	1	Exchange top of stack, H & L	18
SPHL	1	1	1	1	1	0	0	1	H & L to stack pointer	5
LXI SP	0	0	1	1	0	0	0	1	Load immediate stack pointer	10
INX SP	0	0	1	1	0	0	1	1	Increment stack pointer	5
DCX SP	0	0	1	1	1	0	1	1	Decrement stack pointer	5
JUMP										
JMP	1	1	0	0	0	0	1	1	Jump unconditional	10
JC	1	1	0	1	1	0	1	0	Jump on carry	10
JNC	1	1	0	1	0	0	1	0	Jump on no carry	10
JZ	1	1	0	0	1	0	1	0	Jump on zero	10
JNZ	1	1	0	0	0	0	1	0	Jump on no zero	10
JP	1	1	1	1	0	0	1	0	Jump on positive	10
JM	1	1	1	1	1	0	1	0	Jump on minus	10
JPE	1	1	1	0	1	0	1	0	Jump on parity even	10
JPO	1	1	1	0	0	0	1	0	Jump on parity odd	10
PCHL	1	1	1	0	1	0	0	1	H & L to program counter	5
CALL										
CALL	1	1	0	0	1	1	0	1	Call unconditional	17
CC	1	1	0	1	1	1	0	0	Call on carry	11/17
CNC	1	1	0	1	0	1	0	0	Call on no carry	11/17
CZ	1	1	0	0	1	1	0	0	Call on zero	11/17
CNZ	1	1	0	0	0	1	0	0	Call on no zero	11/17
CP	1	1	1	1	0	1	0	0	Call on positive	11/17
CM	1	1	1	1	1	1	0	0	Call on minus	11/17
CPE	1	1	1	0	1	1	0	0	Call on parity even	11/17
CPO	1	1	1	0	0	1	0	0	Call on parity odd	11/17

Figure 6-12: Instruction Set for the 8080 Microprocessor. Courtesy Intel Corporation.

Mnemonic	Instruction Code [1] D7	D6	D5	D4	D3	D2	D1	D0	Operations Description	Clock Cycles [2]
RETURN										
RET	1	1	0	0	1	0	0	1	Return	10
RC	1	1	0	1	1	0	0	0	Return on carry	5/11
RNC	1	1	0	1	0	0	0	0	Return on no carry	5/11
RZ	1	1	0	0	1	0	0	0	Return on zero	5/11
RNZ	1	1	0	0	0	0	0	0	Return on no zero	5/11
RP	1	1	1	1	0	0	0	0	Return on positive	5/11
RM	1	1	1	1	1	0	0	0	Return on minus	5/11
RPE	1	1	1	0	1	0	0	0	Return on parity even	5/11
RPO	1	1	1	0	0	0	0	0	Return on parity odd	5/11
RESTART										
RST	1	1	A	A	A	1	1	1	Restart	11
INCREMENT AND DECREMENT										
INR r	0	0	D	D	D	1	0	0	Increment register	5
DCR r	0	0	D	D	D	1	0	1	Decrement register	5
INR M	0	0	1	1	0	1	0	0	Increment memory	10
DCR M	0	0	1	1	0	1	0	1	Decrement memory	10
INX B	0	0	0	0	0	0	1	1	Increment B & C registers	5
INX D	0	0	0	1	0	0	1	1	Increment D & E registers	5
INX H	0	0	1	0	0	0	1	1	Increment H & L registers	5
DCX B	0	0	0	0	1	0	1	1	Decrement B & C	5
DCX D	0	0	0	1	1	0	1	1	Decrement D & E	5
DCX H	0	0	1	0	1	0	1	1	Decrement H & L	5
ADD										
ADD r	1	0	0	0	0	S	S	S	Add register to A	4
ADC r	1	0	0	0	1	S	S	S	Add register to A with carry	4
ADD M	1	0	0	0	0	1	1	0	Add memory to A	7
ADC M	1	0	0	0	1	1	1	0	Add memory to A with carry	7
ADI	1	1	0	0	0	1	1	0	Add immediate to A	7
ACI	1	1	0	0	1	1	1	0	Add immediate to A with carry	7
DAD B	0	0	0	0	1	0	0	1	Add B & C to H & L	10
DAD D	0	0	0	1	1	0	0	1	Add D & E to H & L	10
DAD H	0	0	1	0	1	0	0	1	Add H & L to H & L	10
DAD SP	0	0	1	1	1	0	0	1	Add stack pointer to H & L	10
SUBTRACT										
SUB r	1	0	0	1	0	S	S	S	Subtract register from A	4
SBB r	1	0	0	1	1	S	S	S	Subtract register from A with borrow	4
SUB M	1	0	0	1	0	1	1	0	Subtract memory from A	7
SBB M	1	0	0	1	1	1	1	0	Subtract memory from A with borrow	7
SUI	1	1	0	1	0	1	1	0	Subtract immediate from A	7
SBI	1	1	0	1	1	1	1	0	Subtract immediate from A with borrow	7
LOGICAL										
ANA r	1	0	1	0	0	S	S	S	And register with A	4
XRA r	1	0	1	0	1	S	S	S	Exclusive Or register with A	4
ORA r	1	0	1	1	0	S	S	S	Or register with A	4
CMP r	1	0	1	1	1	S	S	S	Compare register with A	4
ANA M	1	0	1	0	0	1	1	0	And memory with A	7
XRA M	1	0	1	0	1	1	1	0	Exclusive Or memory with A	7
ORA M	1	0	1	1	0	1	1	0	Or memory with A	7
CMP M	1	0	1	1	1	1	1	0	Compare memory with A	7
ANI	1	1	1	0	0	1	1	0	And immediate with A	7
XRI	1	1	1	0	1	1	1	0	Exclusive Or immediate with A	7
ORI	1	1	1	1	0	1	1	0	Or immediate with A	7
CPI	1	1	1	1	1	1	1	0	Compare immediate with A	7

Figure 6-12 (continued)

Mnemonic	Instruction Code [1] D_7 D_6 D_5 D_4 D_3 D_2 D_1 D_0	Operations Description	Clock Cycles [2]
ROTATE			
RLC	0 0 0 0 0 1 1 1	Rotate A left	4
RRC	0 0 0 0 1 1 1 1	Rotate A right	4
RAL	0 0 0 1 0 1 1 1	Rotate A left through carry	4
RAR	0 0 0 1 1 1 1 1	Rotate A right through carry	4
SPECIALS			
CMA	0 0 1 0 1 1 1 1	Complement A	4
STC	0 0 1 1 0 1 1 1	Set carry	4
CMC	0 0 1 1 1 1 1 1	Complement carry	4
DAA	0 0 1 0 0 1 1 1	Decimal adjust A	4
INPUT/OUTPUT			
IN	1 1 0 1 1 0 1 1	Input	10
OUT	1 1 0 1 0 0 1 1	Output	10
CONTROL			
EI	1 1 1 1 1 0 1 1	Enable Interrupts	4
DI	1 1 1 1 0 0 1 1	Disable Interrupt	4
NOP	0 0 0 0 0 0 0 0	No-operation	4
HLT	0 1 1 1 0 1 1 0	Halt	7

NOTES:

1. DDD or SSS: B=000, C=001, D=010, E=011, H=100, L=101, Memory=110, A=111.
2. Two possible cycle times (6/12) indicate instruction cycles dependent on condition flags.

*All mnemonics copyright ©Intel Corporation 1977

Figure 6-12 (continued)

The 8080 microprocessor needs some external chips in order to operate properly. It needs a crystal oscillator for the clocking functions and it needs a system controller and bus driver. The 8334 I.C. provides the controller and bus driver functions. A complete system incorporating the three is shown in Figure 6-13.

THE 6800 MICROPROCESSOR

The Motorola 6800 is another popular microprocessor. One of the primary advantages of the 6800 is that it requires only one power supply (+5.0V).

A block diagram of the 6800 is shown in Figure 6-14. A summary of the instruction codes is given in Figure 6-15, and the interconnection of a complete 6800 microprocessor system is shown in Figure 6-16. The 6800 system contains peripheral interface adapters (PIA) for interfacing to external equipment. It also contains a synchronous communications interface adapter and both ROM and RAM memory chips.

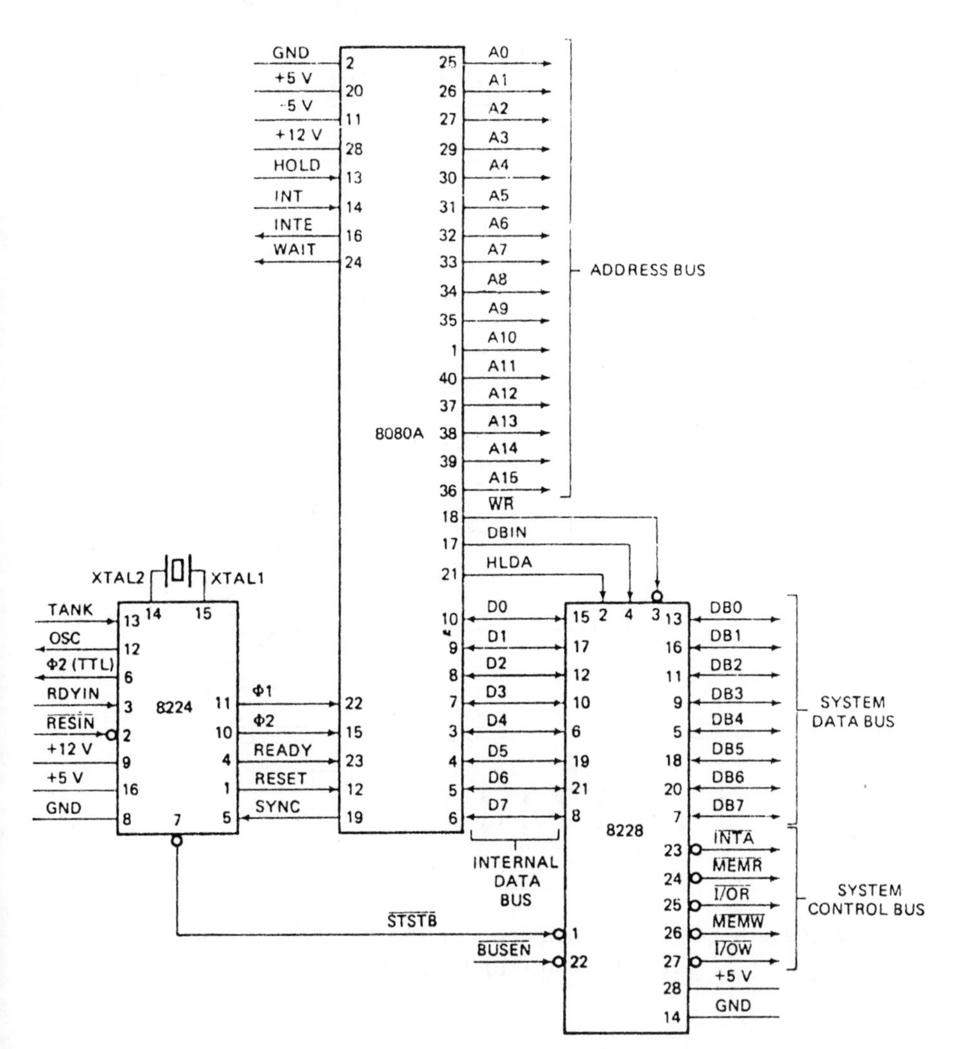

Figure 6-13: An 8080A Microprocessor System. Courtesy Intel Corporation.

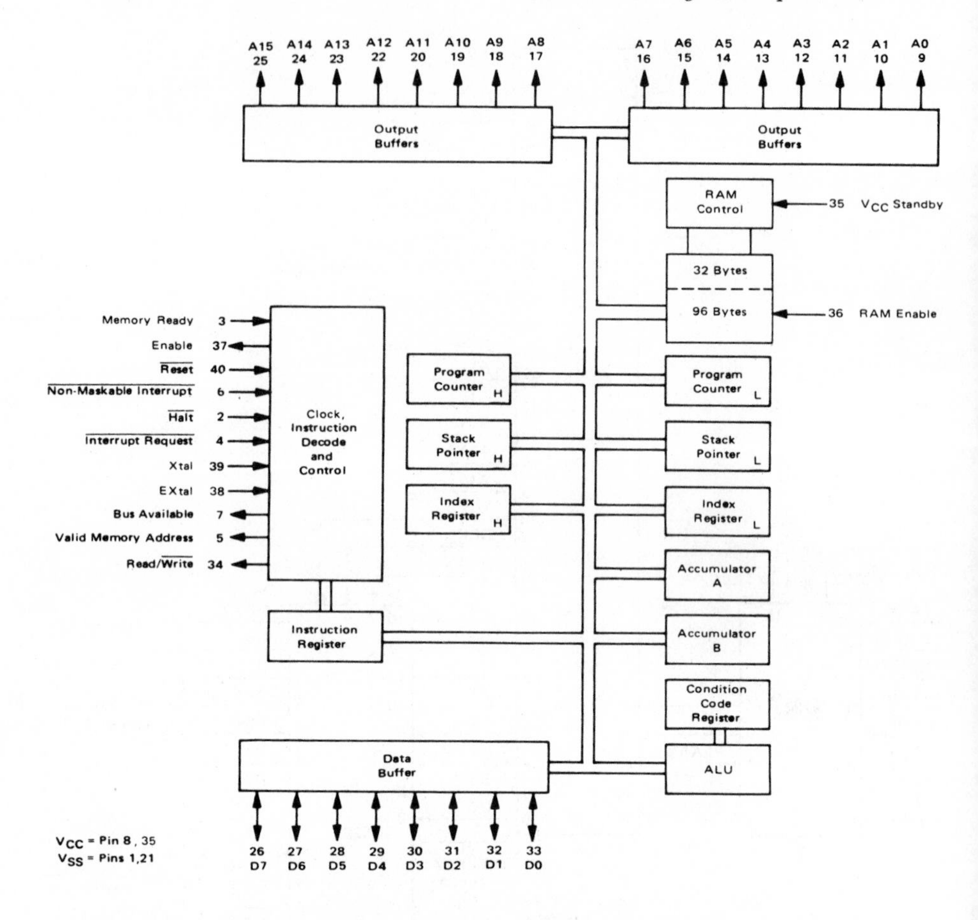

Figure 6-14: Block Diagram of the 6800 Microprocessor. Courtesy of Motorola, Inc.

SUMMARY OF CYCLE BY CYCLE OPERATION

Table 8 provides a detailed description of the information present on the Address Bus, Data Bus, Valid Memory Address line (VMA), and the Read/Write line (R/W) during each cycle for each instruction.

This information is useful in comparing actual with expected results during debug of both software and hardware as the control program is executed. The information is categorized in groups according to Addressing Mode and Number of Cycles per instruction. (In general, instructions with the same Addressing Mode and Number of Cycles execute in the same manner; exceptions are indicated in the table.)

TABLE 8 – OPERATION SUMMARY

Address Mode and Instructions	Cycles	Cycle #	VMA Line	Address Bus	R/W Line	Data Bus
IMMEDIATE						
ADC EOR ADD LDA AND ORA BIT SBC CMP SUB	2	1	1	Op Code Address	1	Op Code
		2	1	Op Code Address + 1	1	Operand Data
CPX LDS LDX	3	1	1	Op Code Address	1	Op Code
		2	1	Op Code Address + 1	1	Operand Data (High Order Byte)
		3	1	Op Code Address + 2	1	Operand Data (Low Order Byte)
DIRECT						
ADC EOR ADD LDA AND ORA BIT SBC CMP SUB	3	1	1	Op Code Address	1	Op Code
		2	1	Op Code Address + 1	1	Address of Operand
		3	1	Address of Operand	1	Operand Data
CPX LDS LDX	4	1	1	Op Code Address	1	Op Code
		2	1	Op Code Address + 1	1	Address of Operand
		3	1	Address of Operand	1	Operand Data (High Order Byte)
		4	1	Operand Address + 1	1	Operand Data (Low Order Byte)
STA	4	1	1	Op Code Address	1	Op Code
		2	1	Op Code Address + 1	1	Destination Address
		3	0	Destination Address	1	Irrelevant Data (Note 1)
		4	1	Destination Address	0	Data from Accumulator
STS STX	5	1	1	Op Code Address	1	Op Code
		2	1	Op Code Address + 1	1	Address of Operand
		3	0	Address of Operand	1	Irrelevant Data (Note 1)
		4	1	Address of Operand	0	Register Data (High Order Byte)
		5	1	Address of Operand + 1	0	Register Data (Low Order Byte)
INDEXED						
JMP	4	1	1	Op Code Address	1	Op Code
		2	1	Op Code Address + 1	1	Offset
		3	0	Index Register	1	Irrelevant Data (Note 1)
		4	0	Index Register Plus Offset (w/o Carry)	1	Irrelevant Data (Note 1)
ADC EOR ADD LDA AND ORA BIT SBC CMP SUB	5	1	1	Op Code Address	1	Op Code
		2	1	Op Code Address + 1	1	Offset
		3	0	Index Register	1	Irrelevant Data (Note 1)
		4	0	Index Register Plus Offset (w/o Carry)	1	Irrelevant Data (Note 1)
		5	1	Index Register Plus Offset	1	Operand Data
CPX LDS LDX	6	1	1	Op Code Address	1	Op Code
		2	1	Op Code Address + 1	1	Offset
		3	0	Index Register	1	Irrelevant Data (Note 1)
		4	0	Index Register Plus Offset (w/o Carry)	1	Irrelevant Data (Note 1)
		5	1	Index Register Plus Offset	1	Operand Data (High Order Byte)
		6	1	Index Register Plus Offset + 1	1	Operand Data (Low Order Byte)

Figure 6-15: 6800 Instruction Codes. Courtesy of Motorola, Inc.

TABLE 8 – OPERATION SUMMARY (Continued)

Address Mode and Instructions	Cycles	Cycle #	VMA Line	Address Bus	R/W Line	Data Bus
INDEXED (Continued)						
STA	6	1	1	Op Code Address	1	Op Code
		2	1	Op Code Address + 1	1	Offset
		3	0	Index Register	1	Irrelevant Data (Note 1)
		4	0	Index Register Plus Offset (w/o Carry)	1	Irrelevant Data (Note 1)
		5	0	Index Register Plus Offset	1	Irrelevant Data (Note 1)
		6	1	Index Register Plus Offset	0	Operand Data
ASL LSR ASR NEG CLR ROL COM ROR DEC TST INC	7	1	1	Op Code Address	1	Op Code
		2	1	Op Code Address + 1	1	Offset
		3	0	Index Register	1	Irrelevant Data (Note 1)
		4	0	Index Register Plus Offset (w/o Carry)	1	Irrelevant Data (Note 1)
		5	1	Index Register Plus Offset	1	Current Operand Data
		6	0	Index Register Plus Offset	1	Irrelevant Data (Note 1)
		7	1/0 (Note 3)	Index Register Plus Offset	0	New Operand Data (Note 3)
STS STX	7	1	1	Op Code Address	1	Op Code
		2	1	Op Code Address + 1	1	Offset
		3	0	Index Register	1	Irrelevant Data (Note 1)
		4	0	Index Register Plus Offset (w/o Carry)	1	Irrelevant Data (Note 1)
		5	0	Index Register Plus Offset	1	Irrelevant Data (Note 1)
		6	1	Index Register Plus Offset	0	Operand Data (High Order Byte)
		7	1	Index Register Plus Offset + 1	0	Operand Data (Low Order Byte)
JSR	8	1	1	Op Code Address	1	Op Code
		2	1	Op Code Address + 1	1	Offset
		3	0	Index Register	1	Irrelevant Data (Note 1)
		4	1	Stack Pointer	0	Return Address (Low Order Byte)
		5	1	Stack Pointer – 1	0	Return Address (High Order Byte)
		6	0	Stack Pointer – 2	1	Irrelevant Data (Note 1)
		7	0	Index Register	1	Irrelevant Data (Note 1)
		8	0	Index Register Plus Offset (w/o Carry)	1	Irrelevant Data (Note 1)
EXTENDED						
JMP	3	1	1	Op Code Address	1	Op Code
		2	1	Op Code Address + 1	1	Jump Address (High Order Byte)
		3	1	Op Code Address + 2	1	Jump Address (Low Order Byte)
ADC EOR ADD LDA AND ORA BIT SBC CMP SUB	4	1	1	Op Code Address	1	Op Code
		2	1	Op Code Address + 1	1	Address of Operand (High Order Byte)
		3	1	Op Code Address + 2	1	Address of Operand (Low Order Byte)
		4	1	Address of Operand	1	Operand Data
CPX LDS LDX	5	1	1	Op Code Address	1	Op Code
		2	1	Op Code Address + 1	1	Address of Operand (High Order Byte)
		3	1	Op Code Address + 2	1	Address of Operand (Low Order Byte)
		4	1	Address of Operand	1	Operand Data (High Order Byte)
		5	1	Address of Operand + 1	1	Operand Data (Low Order Byte)
STA A STA B	5	1	1	Op Code Address	1	Op Code
		2	1	Op Code Address + 1	1	Destination Address (High Order Byte)
		3	1	Op Code Address + 2	1	Destination Address (Low Order Byte)
		4	0	Operand Destination Address	1	Irrelevant Data (Note 1)
		5	1	Operand Destination Address	0	Data from Accumulator
ASL LSR ASR NEG CLR ROL COM ROR DEC TST INC	6	1	1	Op Code Address	1	Op Code
		2	1	Op Code Address + 1	1	Address of Operand (High Order Byte)
		3	1	Op Code Address + 2	1	Address of Operand (Low Order Byte)
		4	1	Address of Operand	1	Current Operand Data
		5	0	Address of Operand	1	Irrelevant Data (Note 1)
		6	1/0 (Note 3)	Address of Operand	0	New Operand Data (Note 3)

Figure 6-15 (continued)

TABLE 8 – OPERATION SUMMARY (Continued)

Address Mode and Instructions	Cycles	Cycle #	VMA Line	Address Bus	R/W Line	Data Bus
EXTENDED (Continued)						
STS STX	6	1	1	Op Code Address	1	Op Code
		2	1	Op Code Address + 1	1	Address of Operand (High Order Byte)
		3	1	Op Code Address + 2	1	Address of Operand (Low Order Byte)
		4	0	Address of Operand	1	Irrelevant Data (Note 1)
		5	1	Address of Operand	0	Operand Data (High Order Byte)
		6	1	Address of Operand + 1	0	Operand Data (Low Order Byte)
JSR	9	1	1	Op Code Address	1	Op Code
		2	1	Op Code Address + 1	1	Address of Subroutine (High Order Byte)
		3	1	Op Code Address + 2	1	Address of Subroutine (Low Order Byte)
		4	1	Subroutine Starting Address	1	Op Code of Next Instruction
		5	1	Stack Pointer	0	Return Address (Low Order Byte)
		6	1	Stack Pointer – 1	0	Return Address (High Order Byte)
		7	0	Stack Pointer – 2	1	Irrelevant Data (Note 1)
		8	0	Op Code Address + 2	1	Irrelevant Data (Note 1)
		9	1	Op Code Address + 2	1	Address of Subroutine (Low Order Byte)
INHERENT						
ABA DAA SEC ASL DEC SEI ASR INC SEV CBA LSR TAB CLC NEG TAP CLI NOP TBA CLR ROL TPA CLV ROR TST COM SBA	2	1	1	Op Code Address	1	Op Code
		2	1	Op Code Address + 1	1	Op Code of Next Instruction
DES DEX INS INX	4	1	1	Op Code Address	1	Op Code
		2	1	Op Code Address + 1	1	Op Code of Next Instruction
		3	0	Previous Register Contents	1	Irrelevant Data (Note 1)
		4	0	New Register Contents	1	Irrelevant Data (Note 1)
PSH	4	1	1	Op Code Address	1	Op Code
		2	1	Op Code Address + 1	1	Op Code of Next Instruction
		3	1	Stack Pointer	0	Accumulator Data
		4	0	Stack Pointer – 1	1	Accumulator Data
PUL	4	1	1	Op Code Address	1	Op Code
		2	1	Op Code Address + 1	1	Op Code of Next Instruction
		3	0	Stack Pointer	1	Irrelevant Data (Note 1)
		4	1	Stack Pointer + 1	1	Operand Data from Stack
TSX	4	1	1	Op Code Address	1	Op Code
		2	1	Op Code Address + 1	1	Op Code of Next Instruction
		3	0	Stack Pointer	1	Irrelevant Data (Note 1)
		4	0	New Index Register	1	Irrelevant Data (Note 1)
TXS	4	1	1	Op Code Address	1	Op Code
		2	1	Op Code Address + 1	1	Op Code of Next Instruction
		3	0	Index Register	1	Irrelevant Data
		4	0	New Stack Pointer	1	Irrelevant Data
RTS	5	1	1	Op Code Address	1	Op Code
		2	1	Op Code Address + 1	1	Irrelevant Data (Note 2)
		3	0	Stack Pointer	1	Irrelevant Data (Note 1)
		4	1	Stack Pointer + 1	1	Address of Next Instruction (High Order Byte)
		5	1	Stack Pointer + 2	1	Address of Next Instruction (Low Order Byte)

Figure 6-15 (continued)

TABLE 8 – OPERATION SUMMARY (Continued)

Address Mode and Instructions	Cycles	Cycle #	VMA Line	Address Bus	R/W Line	Data Bus
INHERENT (Continued)						
WAI	9	1	1	Op Code Address	1	Op Code
		2	1	Op Code Address + 1	1	Op Code of Next Instruction
		3	1	Stack Pointer	0	Return Address (Low Order Byte)
		4	1	Stack Pointer – 1	0	Return Address (High Order Byte)
		5	1	Stack Pointer – 2	0	Index Register (Low Order Byte)
		6	1	Stack Pointer – 3	0	Index Register (High Order Byte)
		7	1	Stack Pointer – 4	0	Contents of Accumulator A
		8	1	Stack Pointer – 5	0	Contents of Accumulator B
		9	1	Stack Pointer – 6 (Note 4)	1	Contents of Cond. Code Register
RTI	10	1	1	Op Code Address	1	Op Code
		2	1	Op Code Address + 1	1	Irrelevant Data (Note 2)
		3	0	Stack Pointer	1	Irrelevant Data (Note 1)
		4	1	Stack Pointer + 1	1	Contents of Cond. Code Register from Stack
		5	1	Stack Pointer + 2	1	Contents of Accumulator B from Stack
		6	1	Stack Pointer + 3	1	Contents of Accumulator A from Stack
		7	1	Stack Pointer + 4	1	Index Register from Stack (High Order Byte)
		8	1	Stack Pointer + 5	1	Index Register from Stack (Low Order Byte)
		9	1	Stack Pointer + 6	1	Next Instruction Address from Stack (High Order Byte)
		10	1	Stack Pointer + 7	1	Next Instruction Address from Stack (Low Order Byte)
SWI	12	1	1	Op Code Address	1	Op Code
		2	1	Op Code Address + 1	1	Irrelevant Data (Note 1)
		3	1	Stack Pointer	0	Return Address (Low Order Byte)
		4	1	Stack Pointer – 1	0	Return Address (High Order Byte)
		5	1	Stack Pointer – 2	0	Index Register (Low Order Byte)
		6	1	Stack Pointer – 3	0	Index Register (High Order Byte)
		7	1	Stack Pointer – 4	0	Contents of Accumulator A
		8	1	Stack Pointer – 5	0	Contents of Accumulator B
		9	1	Stack Pointer – 6	0	Contents of Cond. Code Register
		10	0	Stack Pointer – 7	1	Irrelevant Data (Note 1)
		11	1	Vector Address FFFA (Hex)	1	Address of Subroutine (High Order Byte)
		12	1	Vector Address FFFB (Hex)	1	Address of Subroutine (Low Order Byte)
RELATIVE						
BCC BHI BNE BCS BLE BPL BEQ BLS BRA BGE BLT BVC BGT BMI BVS	4	1	1	Op Code Address	1	Op Code
		2	1	Op Code Address + 1	1	Branch Offset
		3	0	Op Code Address + 2	1	Irrelevant Data (Note 1)
		4	0	Branch Address	1	Irrelevant Data (Note 1)
BSR	8	1	1	Op Code Address	1	Op Code
		2	1	Op Code Address + 1	1	Branch Offset
		3	0	Return Address of Main Program	1	Irrelevant Data (Note 1)
		4	1	Stack Pointer	0	Return Address (Low Order Byte)
		5	1	Stack Pointer – 1	0	Return Address (High Order Byte)
		6	0	Stack Pointer – 2	1	Irrelevant Data (Note 1)
		7	0	Return Address of Main Program	1	Irrelevant Data (Note 1)
		8	0	Subroutine Address	1	Irrelevant Data (Note 1)

Note 1. If device which is addressed during this cycle uses VMA, then the Data Bus will go to the high impedance three-state condition. Depending on bus capacitance, data from the previous cycle may be retained on the Data Bus.
Note 2. Data is ignored by the MPU.
Note 3. For TST, VMA = 0 and Operand data does not change.
Note 4. While the MPU is waiting for the interrupt, Bus Available will go high indicating the following states of the control lines: VMA is low; Address Bus, R/W, and Data Bus are all in the high impedance state.

Figure 6-15 (continued)

MORE MICROPROCESSORS

The evolution of microprocessors has progressed from arithmetic logic circuits (ALUs) and industrial control units (ICUs), to a 4-bit device like the 4004, to 8-bit devices like the 8080 and 6800. This is by no means the end result of microprocessor evolution.

There is one basic way to create better microprocessors and that is to put more circuits on the same chip. These circuits could take the place of some of the external control type of chips necessary for the 8080 or 6800 or these circuits could be used to allow 16-bit data to be manipulated.

Figure 6-13 shows the support chips necessary for the operation of the 8080 microprocessor. It also shows that three supply voltages (+5v), (−5v), (+12v) are needed. This is all simplified in a microprocessor such as the 6802. The 6802 is a Motorola product that is software compatible with the 6800. The 6802 operates from a single (+5v) supply. The clock chip is built into the device and there is an 128 × 8-bit R.A.M. chip built in.

The 6802 can be used as the basis of a microcomputer using a very few chips. Such a microcomputer system is shown in Figure 6-17.

16-bit microprocessors have also become available in the last few years. Since they handle 16 bits of data at a time, they are much more efficient than an 8-bit device. A pin-out of the 68000 (16-bit) microprocessor is shown in Figure 6-18.

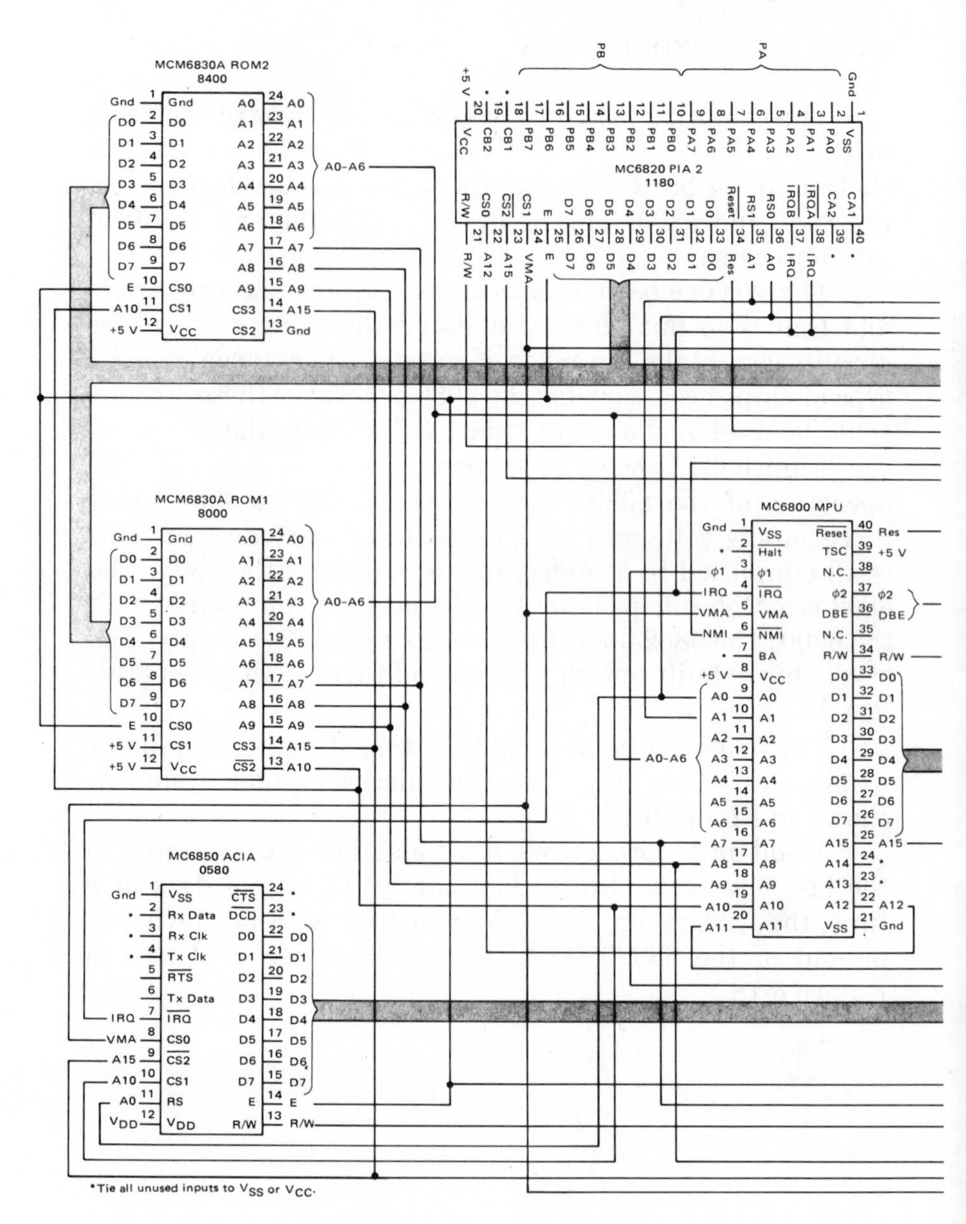

Figure 6-16: A Complete Microprocessor System. Courtesy of Motorola, Inc.

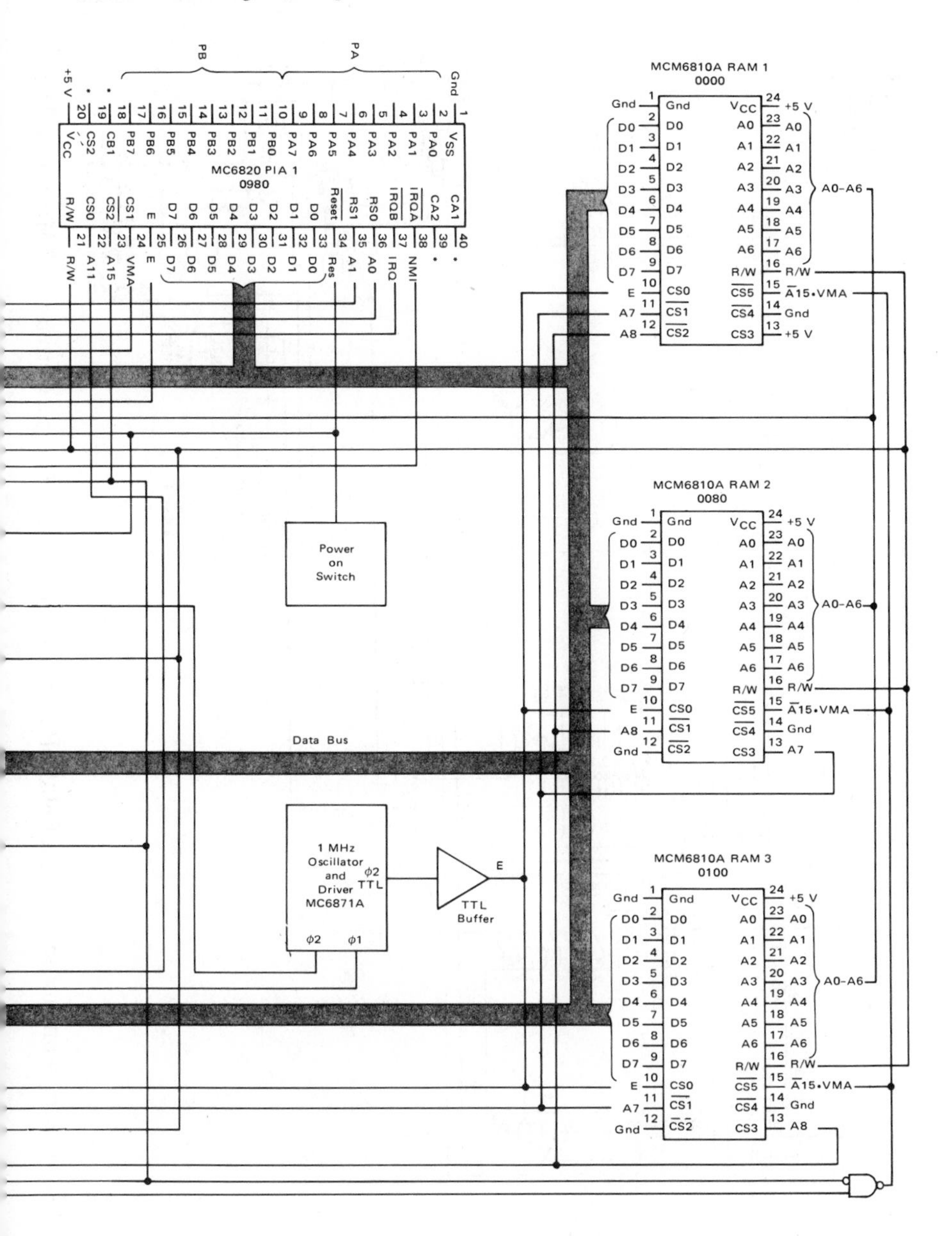
MC6820 PIA 1
0980
MCM6810A RAM 1
0000
MCM6810A RAM 2
0080
MCM6810A RAM 3
0100
Power on Switch
Data Bus
1 MHz Oscillator and Driver MC6871A
TTL Buffer
A0–A6
$\overline{A15}$•VMA

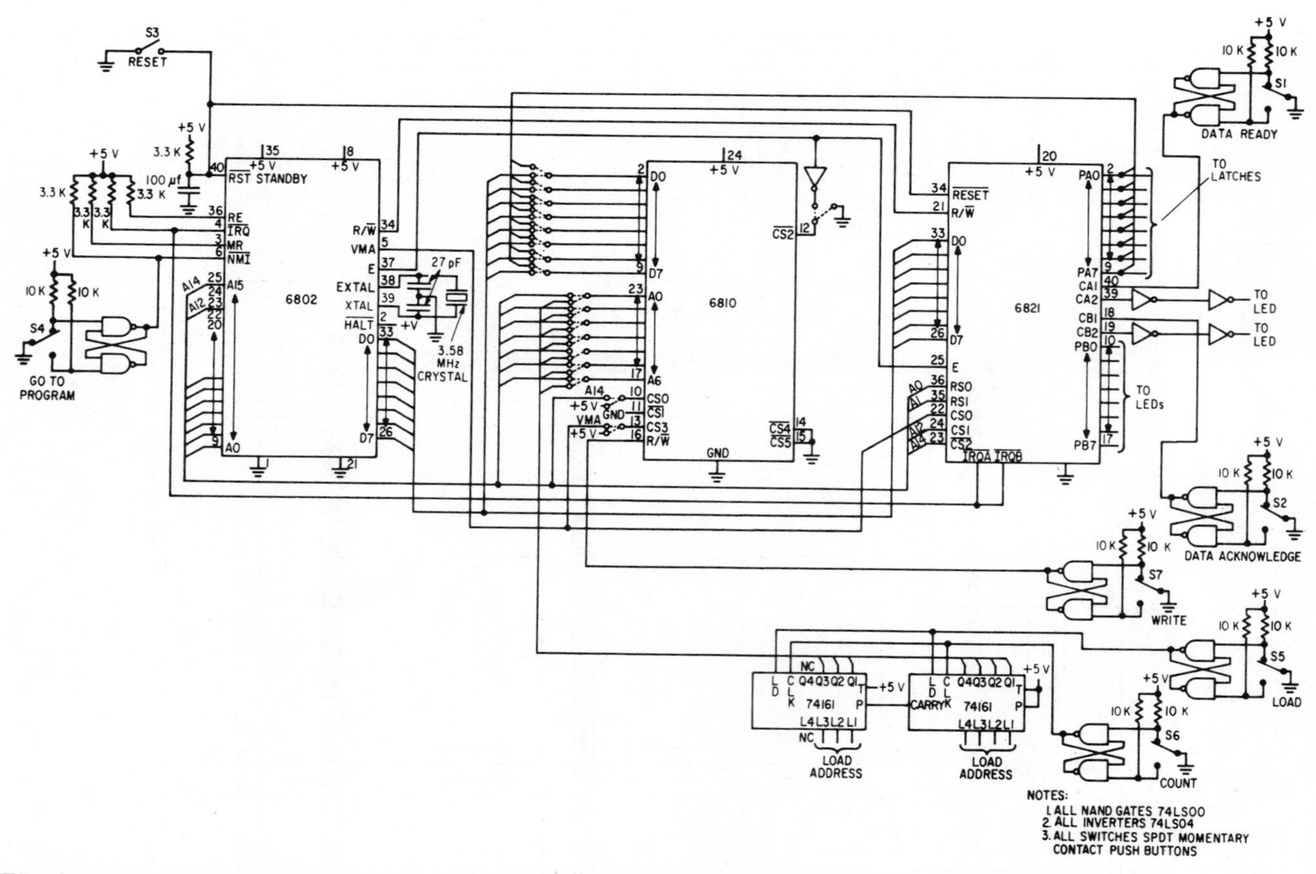

Figure 6-17: A Microcomputer Using a 6802 μP and a Minimum of Chips. Courtesy of Motorola, Inc.

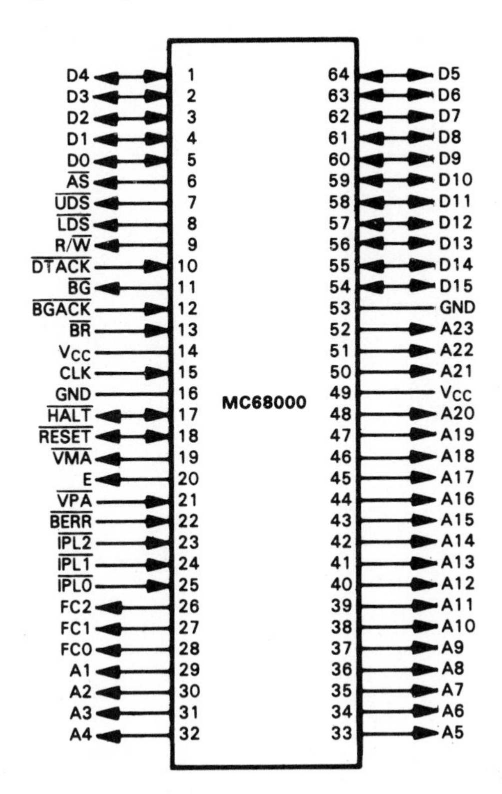

Pin Name	Description	Type
D0-D15	Data Bus	Bidirectional, Tristate
A1-A23	Address Bus	Output, Tristate
$\overline{AS}$	Address Strobe	Output, Tristate
R/$\overline{W}$	Read/Write Control	Output, Tristate
$\overline{UDS}$, $\overline{LDS}$	Upper, Lower Data Strobes	Output, Tristate
$\overline{DTACK}$	Data Transfer Acknowledge	Input
FC0, FC1, FC2	Function Code (status) Outputs	Output, Tristate
$\overline{IPL0}$, $\overline{IPL1}$, $\overline{IPL2}$	Interrupt Requests	Input
$\overline{BERR}$	Bus Error	Input
$\overline{HALT}$	Halt Processor Operation	Input/Output
$\overline{RESET}$	Reset Processor or Reset External Devices	Input/Output
CLK	System Clock	Input
$\overline{BR}$	Bus Request	Input
$\overline{BG}$	Bus Grant	Output
$\overline{BGACK}$	Bus Grant Acknowledge	Input
E	Enable (Clock) Output	Output
$\overline{VMA}$	Valid Memory Address	Output, Tristate
$\overline{VPA}$	Valid Peripheral Address	Input
V_{CC}, GND	Power (+5 V) and Ground	

Figure 6-18: 16-Bit Microprocessor. Courtesy of Motorola, Inc.

7

Explanation of Memory Devices

INTRODUCTION

Computer memory devices have evolved through several types of devices and will probably continue to evolve. The driving force is the increase in storage density and corresponding reduction in cost-per-byte. Current memory devices can be divided into two large categories:

1. Semiconductor devices
2. Magnetic tape and discs.

These two categories are the ones covered in this chapter, although several other memory devices do exist.

MAGNETIC CORE

Core memory consists of small doughnut-shaped pieces of ferrite material. Through the center of the doughnut are write lines and sense lines. The core can be magnetized in two different directions representing ones and zeros. Core memories have to be hand-wired and are difficult and expensive to build for large storage systems. Reading a core memory was a destructive process and special electronics had to be built in order to compensate for this problem.

MAGNETIC DRUM

Magnetic drum memories consist of large cylindrical drums with a magnetic coating on the outside. The drums rotate about an axis through the center. Read/write heads are held close to the surface (one per recording track). Data is recorded by changing the current in the head while the drum is rotating.

Most magnetic drum systems are large and cumbersome. Because of this, they are not seen much anymore.

CHARGE-COUPLED DEVICES (CCD)

CCD or charge-coupled devices are a form of semiconductor memory. A typical cross section is shown in Figure 7-1. Storage of binary bits is done in the capacitance between the metal plates and the silicon substrate. The silicon dioxide serves as an insulator and the capacitor dielectric.

The capacitor charge is not permanent and will leak off in some period of time. Because of this, the CCD storage device must be refreshed at some periodic interval.

The physical size of each storage cell of a CCD is quite small compared to a bipolar or even a MOS device. A CCD can store much more data per silicon chip than standard dynamic RAMs.

BUBBLE MEMORY

Bubble memories are made from a special magnetic material called garnet. This garnet is applied as a thin film on a non-magnetic substrate. When the garnet has no external magnetic field present, it has half of its magnetic domains with magnetic north pointing toward the substrate and the other half pointing away from the substrate. When an external field is applied, the garnet's magnetic domains with the opposite polarity condense and form small magnetic bubbles.

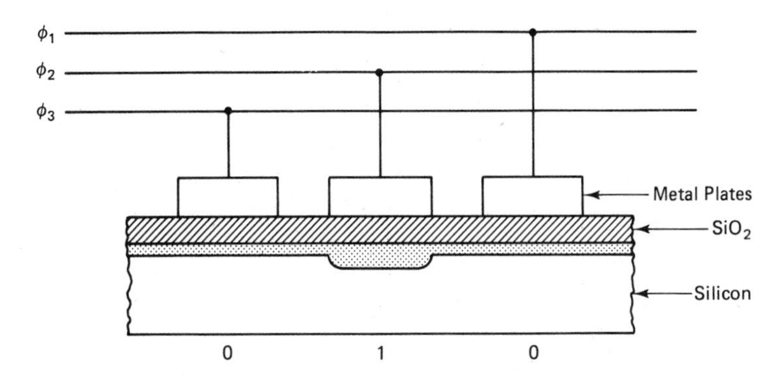

Figure 7-1: Cross Section of C.C.D. Memory.

It is these magnetic bubbles which store data. Presence of a bubble represents a binary 1 and absence of a bubble is a 0. The size of these bubbles is so small that large amounts of data can be stored per chip. The devices are made complex by the circuitry needed to control and move the bubbles.

THE FUNDAMENTALS OF SEMICONDUCTOR MEMORY

With only a few exceptions, semiconductor memories are based on the flip-flop as the basic storage device. The flip-flop contains all the necessary ingredients for storage. It has two stable states to represent ones and zeros. Its state can be easily changed so that it can be written. The output can be sensed (read) without destroying the state of the flip-flop.

Semiconductor memories can be grouped into a family tree as shown in Figure 7-2. It consists of random access memory (RAM) that has read/write capability and read only memory (ROM) from which data can only be read. This family tree further breaks down into bipolar and MOS devices. In order to understand the operation of these devices, it is necessary to understand the operation of the basic flip-flop.

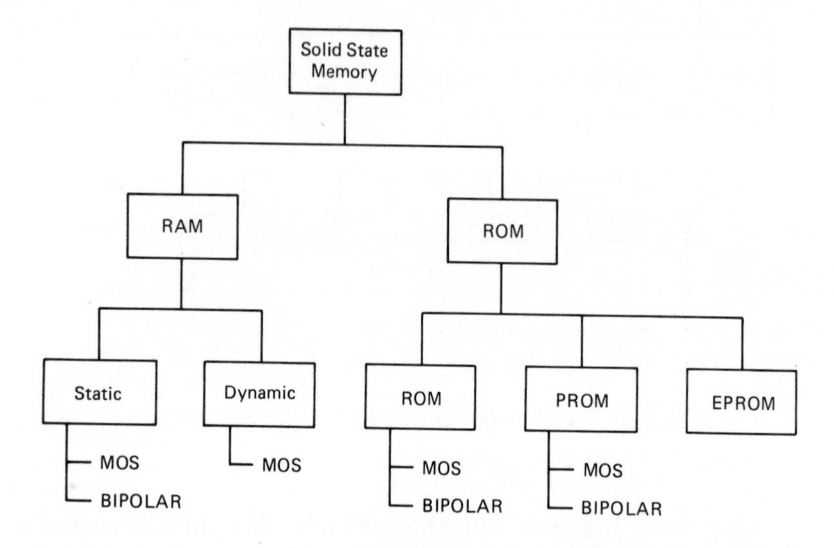

Figure 7-2

BIPOLAR RAM

The basic unit of bipolar RAM is a flip-flop made from bipolar transistors as shown in Figure 7-3. The flip-flops have inputs for addressing (x & y) and input/output lines (1 & 0). The operation of this flip-flop is illustrated as follows:

1. Assume that transistor Q_1 is conducting and therefore transistor Q_2 is not.
2. The current through R_1 produces an output voltage representing a one.
3. There is no current through R_2 representing a zero.
4. The flip-flop contains a one.
5. To change states or write a zero into this flip-flop the x, y address lines must be placed high (binary one) so that the transistor current cannot flow through these lines.

   In order to write a zero, $W_0 = 1$ and $W_1 = 0$ (Figure 7-4). Going through inverters, this places a high across R_1 and a low across R_2. Transistor Q_1 can no longer conduct since all three emitters are high.

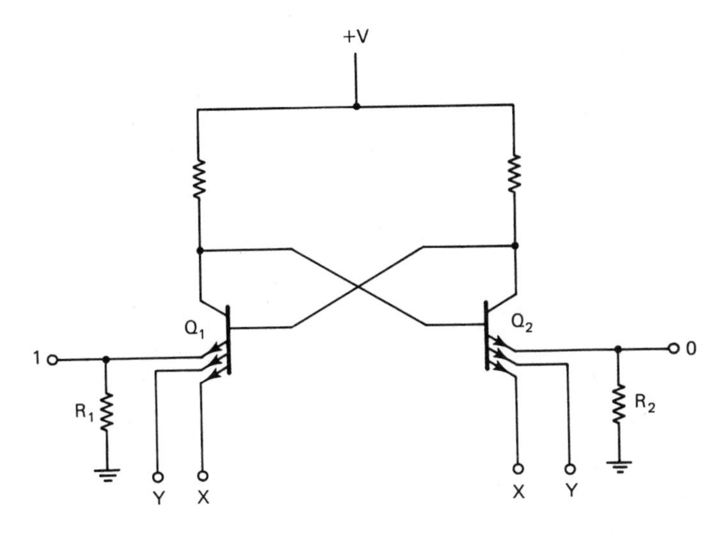

Figure 7-3: Basic Cell for Bipolar R.A.M.

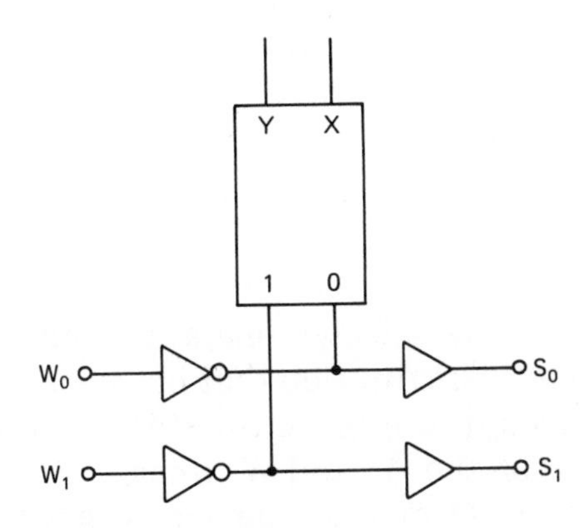

Figure 7-4: R.A.M. Flip-Flop with Write and Sense.

The flip-flop changes states; Q_2 conducts and the flip-flop now contains a zero.

The single flip-flop can only store one binary bit but it is possible to group flip-flops together in order to store large amounts of data. Figure 7-5 shows 16 flip-flops together to form a 16-word by one-bit RAM.

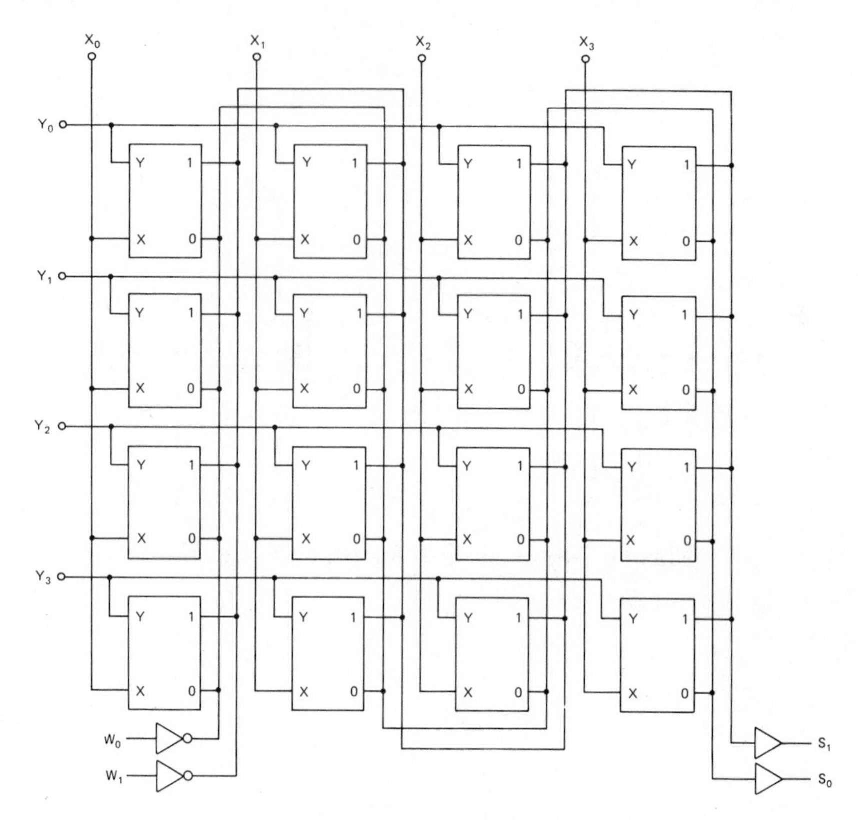

Figure 7-5: A Sixteen-Word by One-Bit R.A.M.

The flip-flops are selected one at a time by addressing the x and y inputs. The flip-flop that has been addressed can now be read through the sense lines or written by placing the proper input on the W_0 and W_1 lines.

A practical RAM memory device is shown in Figure 7-6. All the circuitry shown inside the dotted line exists on one chip. The address lines consist of four inputs for y select and four for x select. These four inputs are decoded by 4 to 16 decoders to address the 16×16 flip-flop array. The chip also contains sense amplifiers, write amplifiers, a chip select input, and a read/write input.

This basic circuitry is used to construct bipolar RAMs with storage capacities of 16 kilobytes and 64 kilobytes.

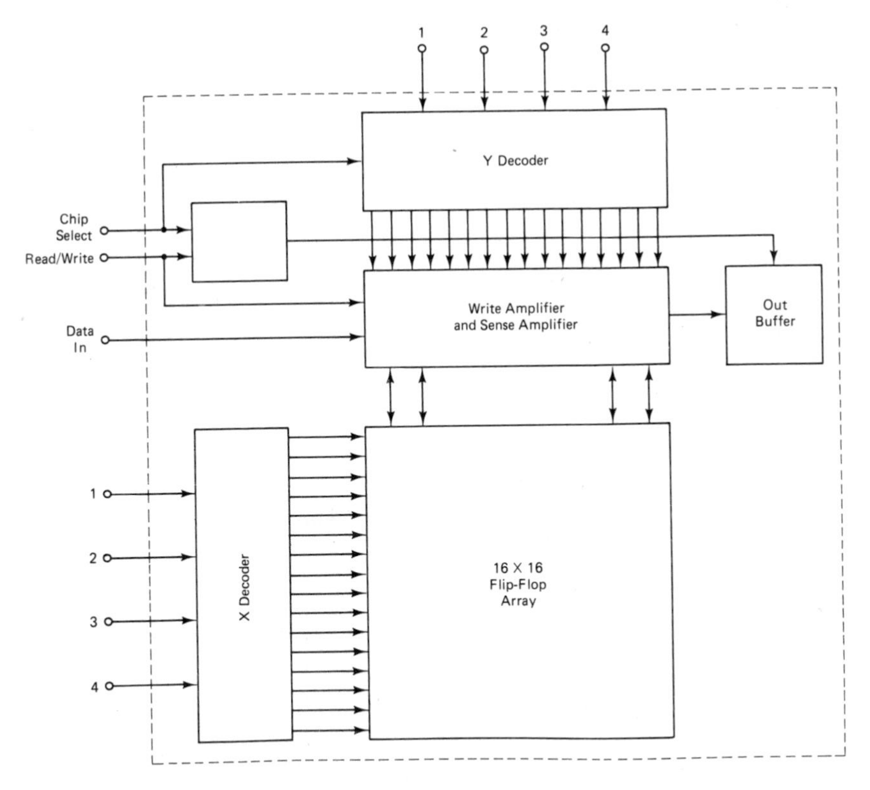

Figure 7-6: A Practical R.A.M. Memory.

MOS STATIC RAMs

Large flip-flop arrays can also be constructed from MOS devices. The basic chip construction shown in Figure 7-6 remains the same but the individual flip-flop cells are made out of MOSFETs instead of bipolar transistors. The basic cell is shown in Figures 7-7 and 7-8; one is a p-channel MOS device and the other is CMOS.

MOS RAMs are divided into two main categories, static and dynamic. Figures 7-7 and 7-8 are static devices. The one bit data is stored in the state of the flip-flop, and unless it is rewritten, this one-bit data will remain indefinitely. Dynamic RAMs store one-bit data as the charge across a

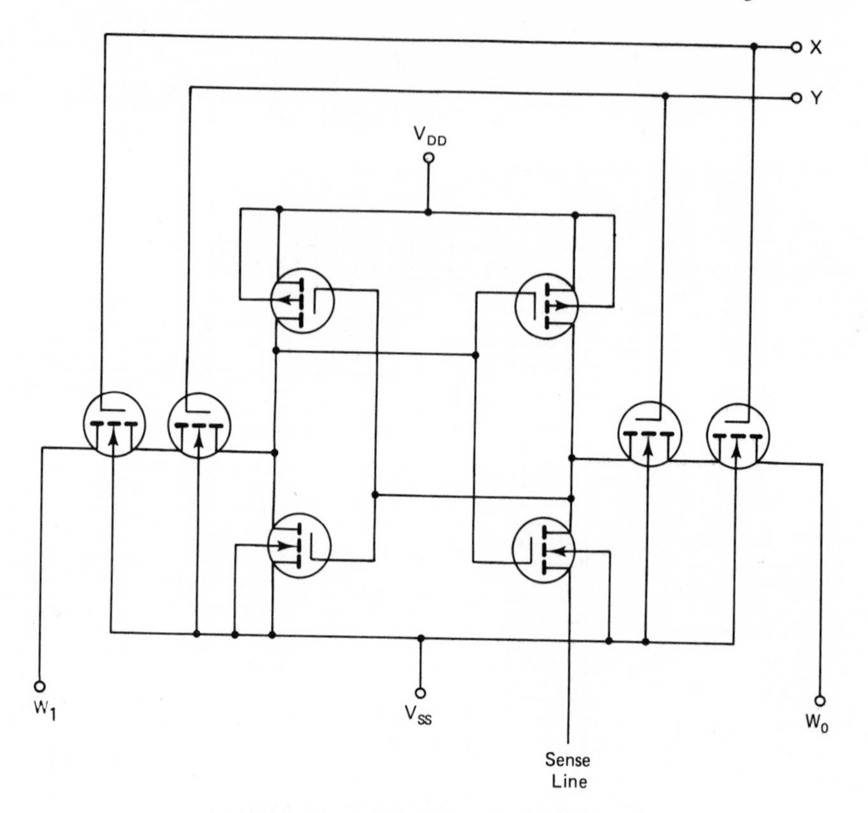

Figure 7-7: CMOS R.A.M. Cell.

capacitor. This capacitance is usually the stray capacitance that exists between the gate and source.

MOS DYNAMIC RAMs

Two different types of basic cells for dynamic RAMs are shown in Figure 7-9(A) and (B). The storage element is the capacitance between the gate and the source. Dynamic RAMs are quite popular in higher density devices such as 64K chips, and they have both advantages and disadvantages:

A. The basic cell contains fewer devices than the static device. This allows much higher packing densities on the chip. For this reason, most 16K and 64K devices are dynamic RAMs.

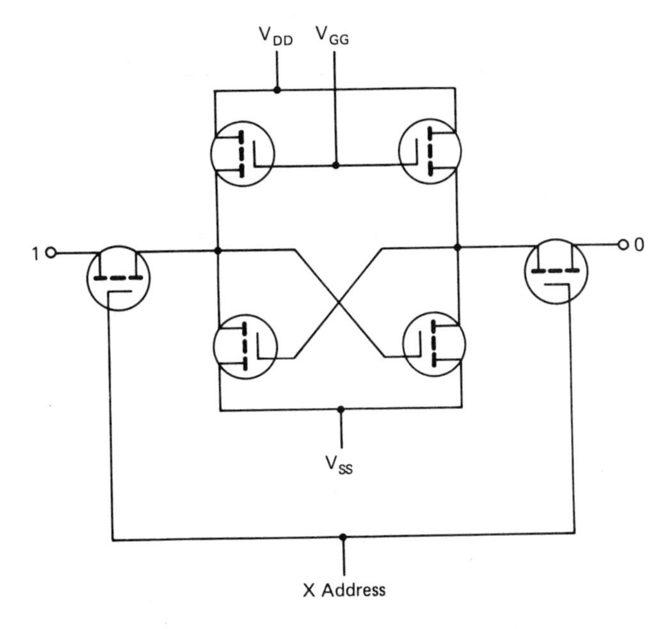

Figure 7-8: P-Channel R.A.M. Cell.

B. Since the basic storage mechanism is charge on a capacitor, the memory is not permanent. The capacitor slowly discharges. For this reason dynamic RAMs must have external circuits used to refresh the memory at periodic intervals. These refresh circuits also contain error detection and correction devices which prevent errors in the memory. A typical circuit for this type of memory is shown in Figure 10-16.

BIPOLAR ROMs

When it is not necessary to change the data in the memory, it is more economical to use a read-only-memory or ROM. Removing the write capability simplifies the basic cell which is shown in Figure 7-10.

The base of the transistor in Figure 7-10 is driven by the address decoder. A one or zero is stored in the cell, depending on the value of R. If R is low ($100\ \Omega - 200\ \Omega$) a zero is stored, if R is an open circuit, a one is stored.

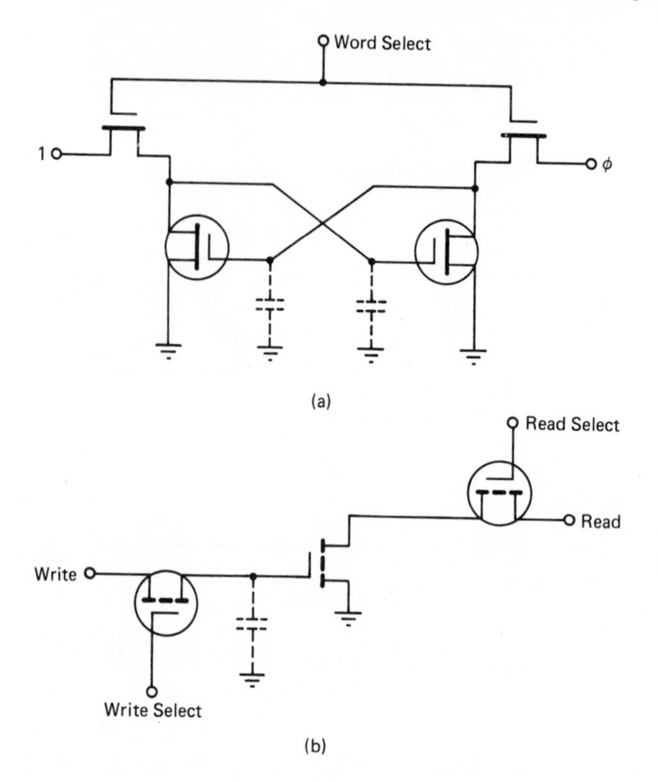

Figure 7-9: Basic Dynamic R.A.M. Cells. A Four Transistor; B Three Transistor.

The ROM is programmed in one of two ways. The first way occurs when the device is manufactured. The resistor is made from aluminum and if a zero is to be stored, the connection is made; if a one is to be stored, the connection is not made. The second method is to manufacture the device, using nichrome as the resistor for each cell. If a one is to be recorded in a cell, a high current pulse is used to open-circuit the nichrome resistor.

The latter device is field programmable. With the proper equipment, the user can create his own programmed ROM. This is a one-time-only programming and if a mistake is made, the device is useless. Devices that can be programmed in this manner are known as PROMs, or programmable read-only-memory.

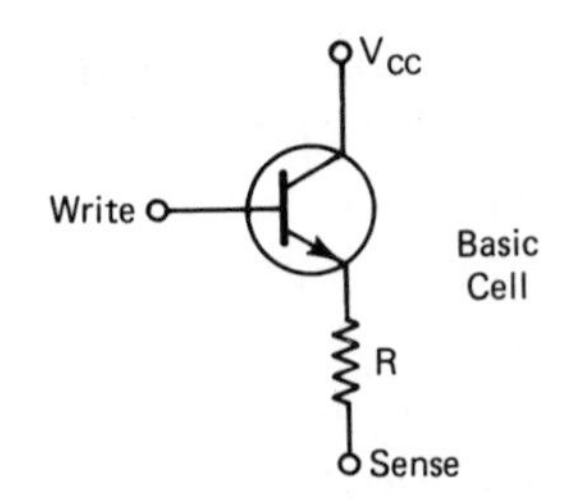

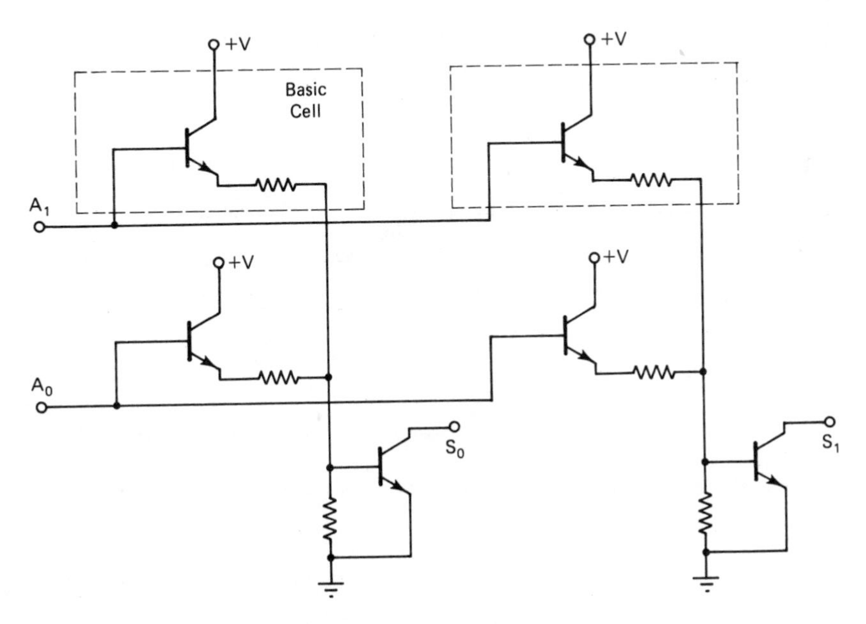

Figure 7-10: Bipolar R.O.M.'s.

Another device in this family is the EPROM or erasable programmable read-only-memory. The EPROM stores binary data in the form of electric charge. It can be erased by the use of ultra-violet light shown through a transparent window on the top of the chip. With the proper equipment, EPROMs can be erased and re-programmed many times.

MOS ROMs

MOS ROMs have the advantage that more active devices can be placed on a given piece of silicon than can be placed using bipolar transistors. The bipolar ROM used a resistor that was conducting or open to represent a zero or a one. The MOS ROM uses the thickness of the oxide layer between the metal gate and the channel. If the oxide is thin, the gate can turn the channel on and off, if the oxide is thick the gate has no effect on the channel. Ones and zeros can then be fabricated into the device during manufacture. A typical cross section of MOS ROM active and inactive transistors is shown in Figure 7-11.

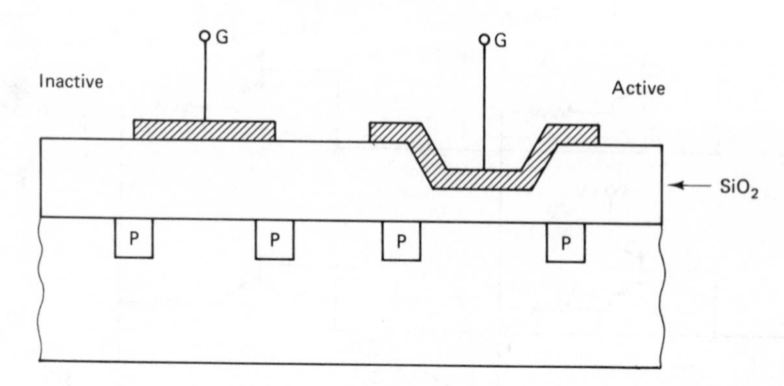

Figure 7-11: Active and Inactive Transistor in a MOS R.O.M.

UNDERSTANDING MAGNETIC RECORDING

At the present time almost all of the mass data storage systems use magnetic recording as the storage vehicle. The cost per megabyte of data storage is 50 to 100 times more economical with magnetic disc and tape systems than with solid state devices.

Magnetic materials can be magnetized in two directions. Therefore, they can easily be made to represent binary ones and binary zeros. The materials used are such that once magnetized they remain magnetized in the same direction until sufficient current is used to change magnetic direction. This allows for mass data storage for indefinite periods of time.

The magnetic flux reversals are placed on the surface through the use of a recording head. The construction of this head is shown in Figure 7-12. It consists of a ferrite material with a gap on the side that interfaces with the recording surface. Around this gap the magnetic flux field extends out beyond the head. It is this flux field that magnetizes and demagnetizes the recording surface. The polarization of these flux lines can be changed by changing the direction of the current through the winding. It is the flux reversals that are used to record binary data.

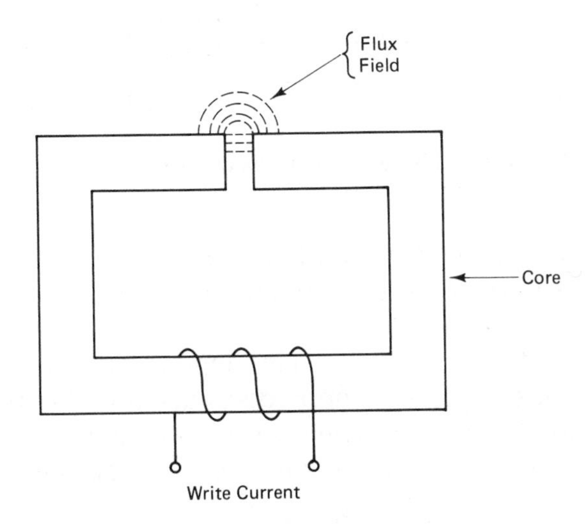

Figure 7-12: Magnetic Recording Head.

In most systems, but not all, the same head and gap used to record (or write) on the magnetic surface is used to read the data. The magnetic flux, once written on the storage media, will produce current in the core winding when the head is passed over the recorded data. A complete read/write system is shown in Figure 7-13. The write amplifier supplies the current necessary to record. During the read operation, the current induced in the coil is amplified by the read amplifier and the signal is then converted into digital pulses.

The recording material itself takes on many different forms. The magnetic material is usually iron oxide, Fe_2O_3.

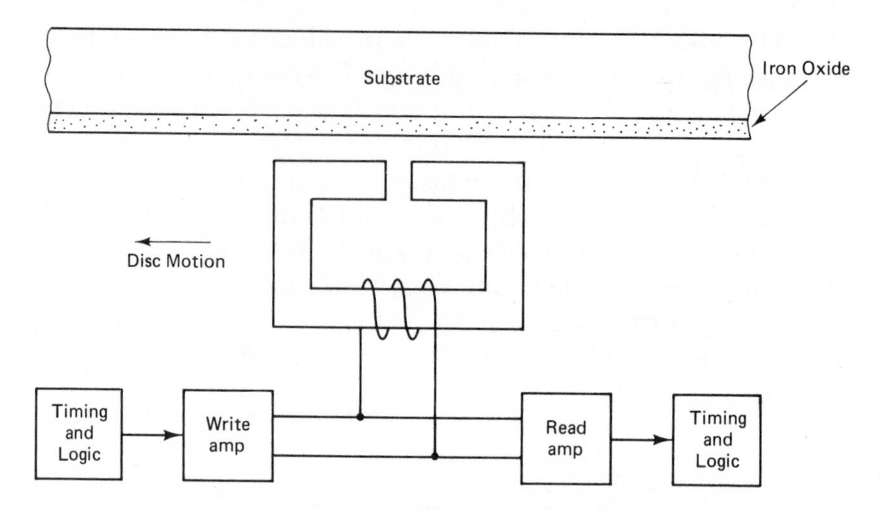

Figure 7-13: Magnetic Recording System.

The oxide comes in very small particles which are oblong in shape; see Figure 7-14, the particles are held together by an epoxy-type binder.

The substrate material varies depending on the application. Magnetic tape and floppy discs use a mylar plastic as the substrate. This gives a flexible substrate so that it can be twisted and bent with no harmful effects. Large magnetic

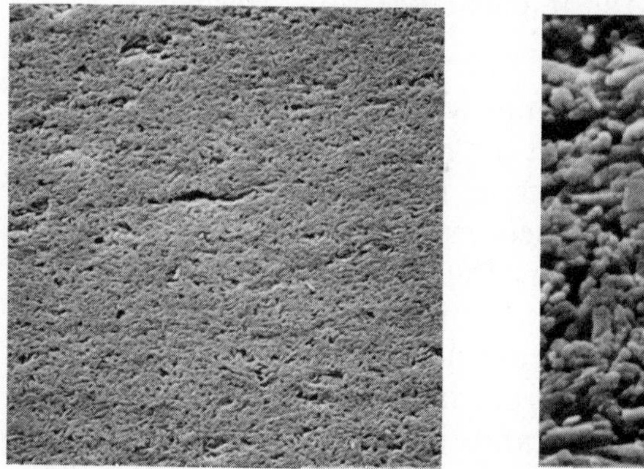

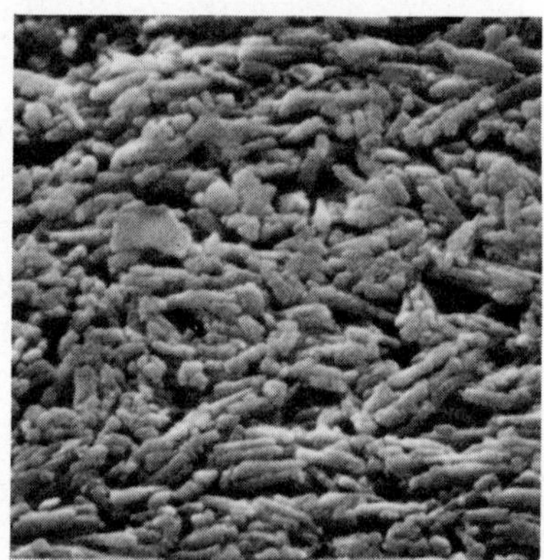

Figure 7-14: Magnetic Gamma Ferric Oxide Particles Magnified 2500 Times and 20,000 Times.

disc storage devices use aluminum as the substrate. These are the so-called rigid disc storage systems. Modern rigid disc systems are capable of storing one to two gigabytes of data.

Whether the system uses tape, floppy disc, or rigid disc, it is common practice in digital recording to saturate the magnetic material. If the write current is increased from zero, the amplitude of the readback signal will increase. This will continue until the surface is saturated. The recording surface at this point is completely magnetized and further increases in current will not change the amplitude. The write current used in these systems is set to be just past the saturation point. A typical saturation curve is shown in Figure 7-15. The typical operating point is shown with a dot.

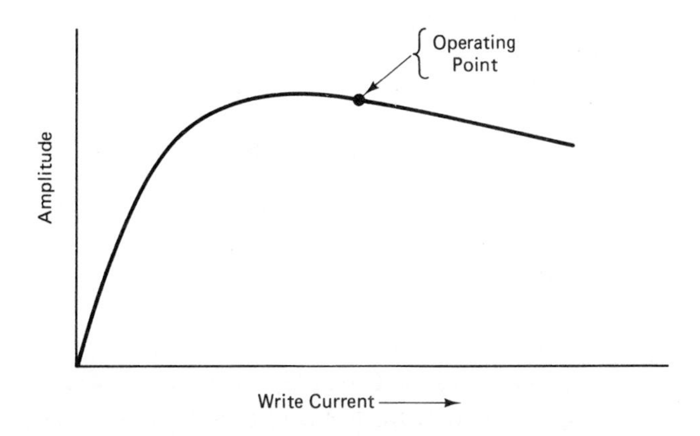

Figure 7-15: Magnetic Saturation Curve.

Figure 7-16 shows typical magnetic flux reversals on the surface along with the write current reversals which caused them. On high-density disc and tape systems, these flux reversals may be only 200 micro-inches apart. The readback signal is also shown in the figure. The output is highest in the area of strong flux reversals since the output must be proportional to d/dT or proportional to the time rate of change of flux.

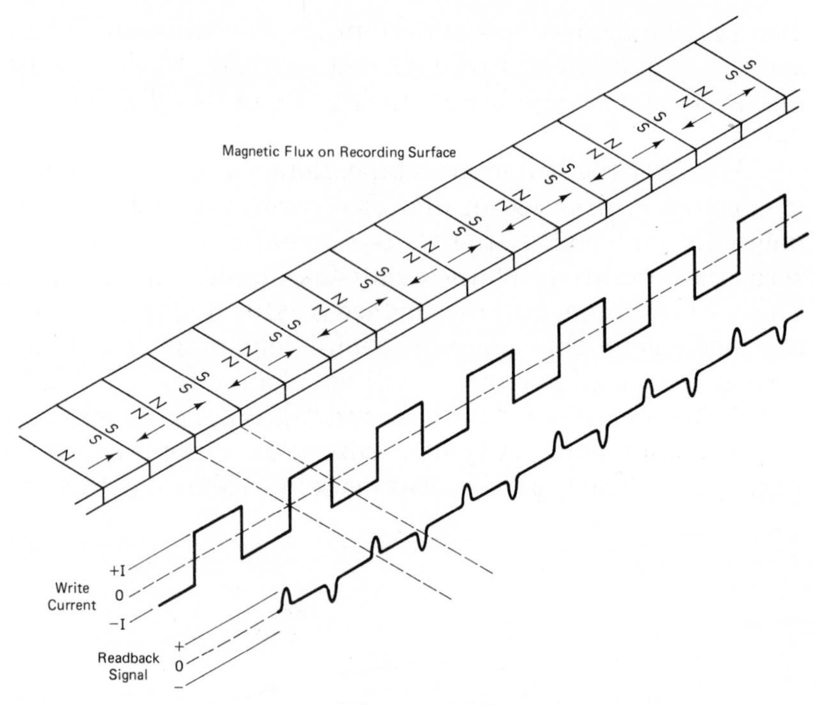

Figure 7-16

MAGNETIC TAPE

Tape storage devices can be divided into two categories: (1) cassettes for small computer systems and (2) seven- and nine-track, high-speed tape for large computer systems.

Cassettes

Most small computers, especially those in the home computer category, can use a standard audio cassette recorder as a data storage device. This is quite different from most digital data recording since the data is stored as audio tones in a non-saturated mode. The audio cassette recorder needs an interface as shown in Figure 7-17. The interface converts the digital data into two audio tones, one tone representing a binary one and another tone representing a binary zero. The interface must also convert the tones back into digital data.

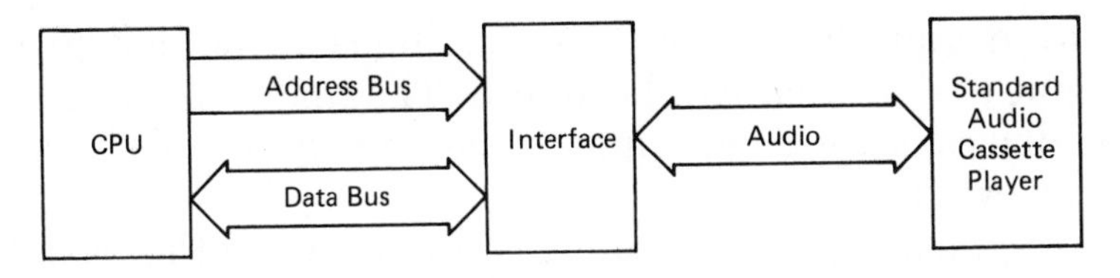

Figure 7-17: System for Using Audio Cassettes.

The frequency, duration, and rate of the tones recorded on cassettes vary depending on the system, and it is usually necessary to consult the appropriate technical manual to see what is being used. There has been at least one attempt to standardize these parameters. Figure 7-18 shows this standard. Eight cycles of a 2400 Hz tone represent a binary one and four cycles of a 1200 Hz tone represent a binary zero. This standard allows a data transfer rate of 300 baud.

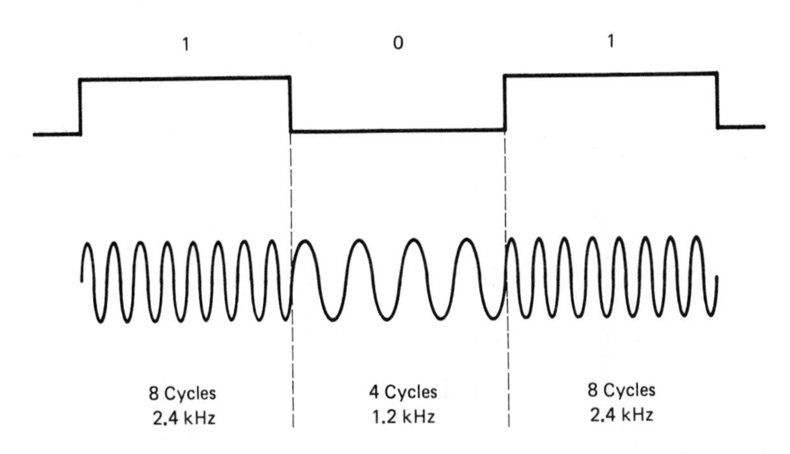

Figure 7-18: Audio Cassette Standard.

It is also common to see systems use these same two frequencies but reduce the number of cycles per bit. This allows the use of higher baud rates.

Large Tape Systems

Tape systems can store massive amounts of data. One reel of tape is very long, and the fact that the reels can be removed and changed means that there is no end to the amount of data that can be stored. The major drawback of tape drives is the access time or the amount of time it takes to retrieve the data.

A typical high-speed tape drive is shown in Figure 7-19. These drives have several items unique to tape drives:

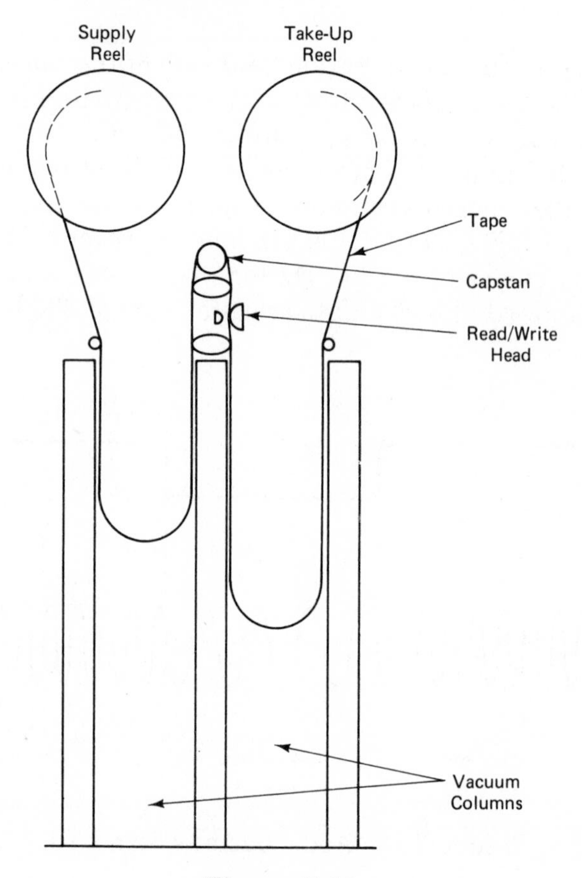

Figure 7-19

A. The read/write head is usually a two-gap head as shown in Figure 7-20. Since the read gap is positioned behind the write gap, it is possible to read the data immediately after writing. This allows fast checking of the data written.

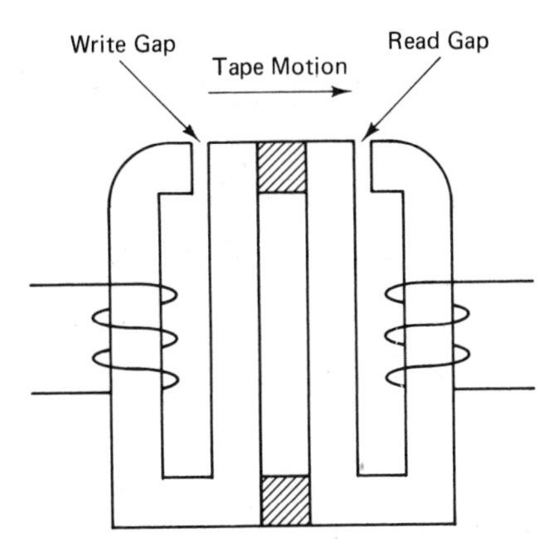

Figure 7-20

B. Since most data on tape is recorded in blocks, it is necessary to start and stop the tape very fast; fast being on the order of 1 millisecond. The large vacuum columns serve to isolate the reels from the capstan and allow fast starting/stopping without harm to the tape.

C. Relative velocity between head and tape is very high, somewhere between 50 and 200 inches per second. While most magnetic recording is considered "contact" recording the high-speed tapes literally fly past the head. The separation can be several micro-inches.

D. Tapes have metallic sections that mark the beginning of the tape and the end of the tape. (BOT & EOT). These metal strips are detected by optical methods.

E. Most tapes used in large systems are one-half inch wide and 0.0015 inches thick. They are recorded in seven tracks or nine tracks. The formats used on seven-track and nine-track tape are shown in Figure 7-21.

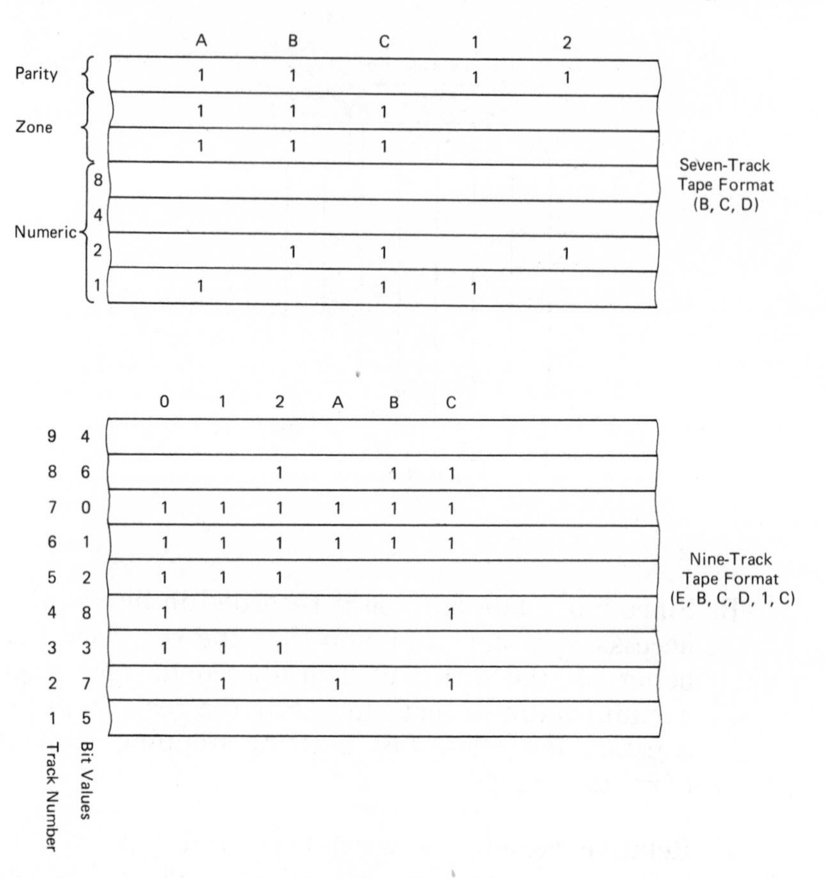

Figure 7-21

MAGNETIC DISCS

Magnetic disc systems can be divided into two large categories, (1) the floppy disc and (2) the rigid disc.

Floppy Discs

Floppy discs are made with a thin magnetic coating on a mylar substrate. They come in two basic sizes, the 8-inch disc and the minifloppy or $5\frac{1}{4}$-inch disc. They are arranged as shown in Figure 7-22. Formats, encoding schemes, and even the number of tracks vary from system to system, so it

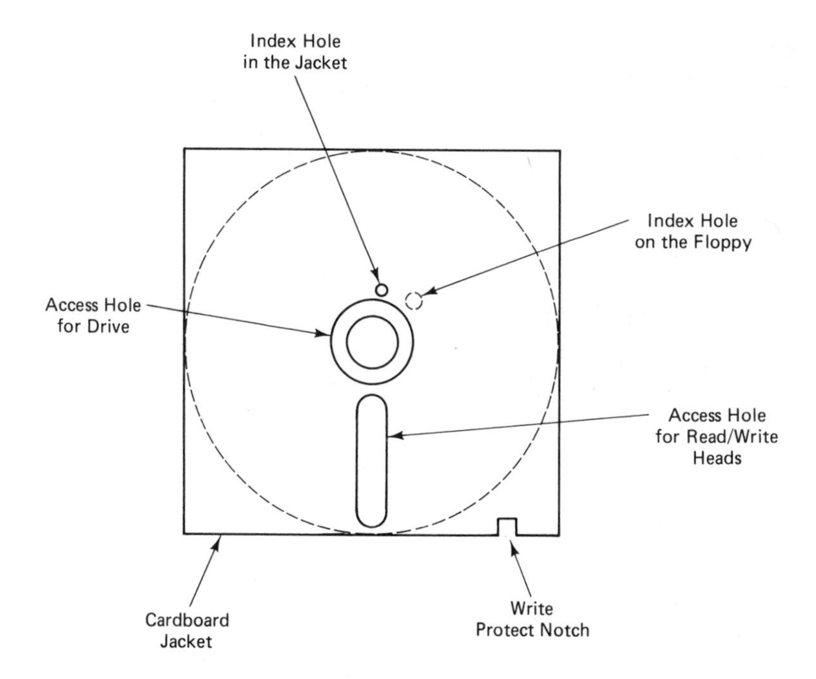

Figure 7-22: Floppy Disc.

is necessary to check the manual to determine the specifics of any floppy disc system.

Rigid Discs

Rigid discs are made with a thin magnetic coating on an aluminum substrate. Most of the rigid discs are 0.075 inches thick (some older products are 0.050) and they come in 14-, 8-, and $5\frac{1}{4}$-inch diameters. Large rigid disc drives are capable of storing one to two gigabytes of data with average data access times of 16 milliseconds. Rigid disc storage systems come with removable cartridges, removable packs, or with the more recent Winchester technology head disc assemblies (HDAs). Winchester technology refers to disc storage devices where the heads and discs are assembled as one unit and the discs are lubricated to enhance reliability. Disc drives can record up to 1,000 tracks per inch and 8,000 bits per inch.

EXAMPLE OF DATA ENCODING

There have been several schemes developed to write binary data on magnetic recording surfaces. The ones in common usage have advantages in one of the two following areas:

1. The ratio of binary bits to magnetic flux reversals should be as high as possible. This allows greater storage densities since a limiting factor in magnetic storage is how close the flux changes can be placed on the media.
2. The ability of the data signal to be self-clocking. If the signal is not self-clocking, some external system of clocking must be used.

One of the first data encoding schemes is called return-to-zero. An example of return-to-zero is shown in Figure 7-23. The write current is in one direction when a binary one is being recorded, in the opposite direction when a zero is being recorded, and in between bits the write current is zero, hence the name return-to-zero.

From Figure 7-23 it can be seen that RZ has two flux changes for each binary bit. It is a simple encoding scheme but not a very efficient one. There exist numerous variations of the basic RZ scheme, each designed for specific applications, but the most commonly used variation of the RZ encoding scheme is called non-return-to-zero-inverted (NRZI).

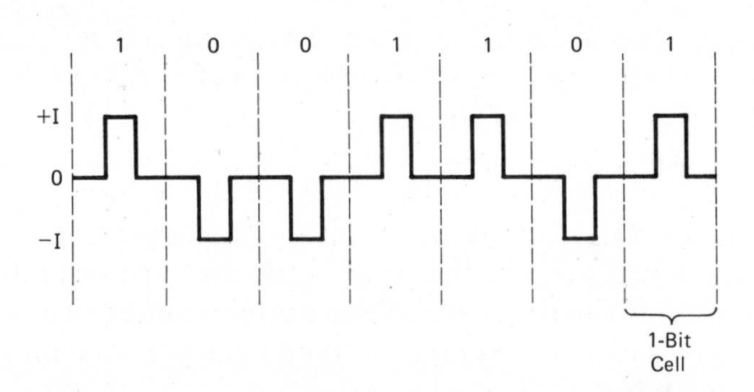

Figure 7-23: RZ (Return-to-Zero) Data Encoding.

NRZI

The NRZI encoding scheme is shown in Figure 7-24. For every binary 1, there is a corresponding magnetic flux reversal on the medium. For every binary 0, there is no flux reversal.

The NRZI encoding method has an efficiency of 1.0, since there is a flux reversal for each binary 1. The highest density data pattern is an all 1's data pattern. The lowest density pattern is an all 0's pattern. The all 0's pattern produces no flux reversals on the medium; instead the medium is magnetized all in the same direction.

NRZI is not a self-clocking code. The frequencies involved are totally a function of the data pattern being recorded. However, it is very easy to encode. Figure 7-25 shows the pulses for the binary data pattern 1110101. To encode this data in NRZI requires only a divide-by-two flip-flop. This produces the required current change for each binary one.

FM

The FM (frequency modulation) encoding scheme is shown in Figure 7-26. There is a flux reversal at the boundary of each bit cell. If the cell contains a one, there is a flux reversal in the center of the cell. If the cell contains a zero, there is no flux reversal in the center of the cell.

The FM scheme has very poor efficiency because of the flux reversals at the cell boundaries whether or not the cell contains a one. The code does have the advantage that it is self-clocking. The flux reversals at the cell boundaries provide the timing window in which a 1 can exist or not exist.

MFM

The MFM (modified frequency modulation) encoding scheme is shown in Figure 7-27. The MFM scheme is based on the FM scheme but provides a much better efficiency. It still maintains the self-clocking ability of FM.

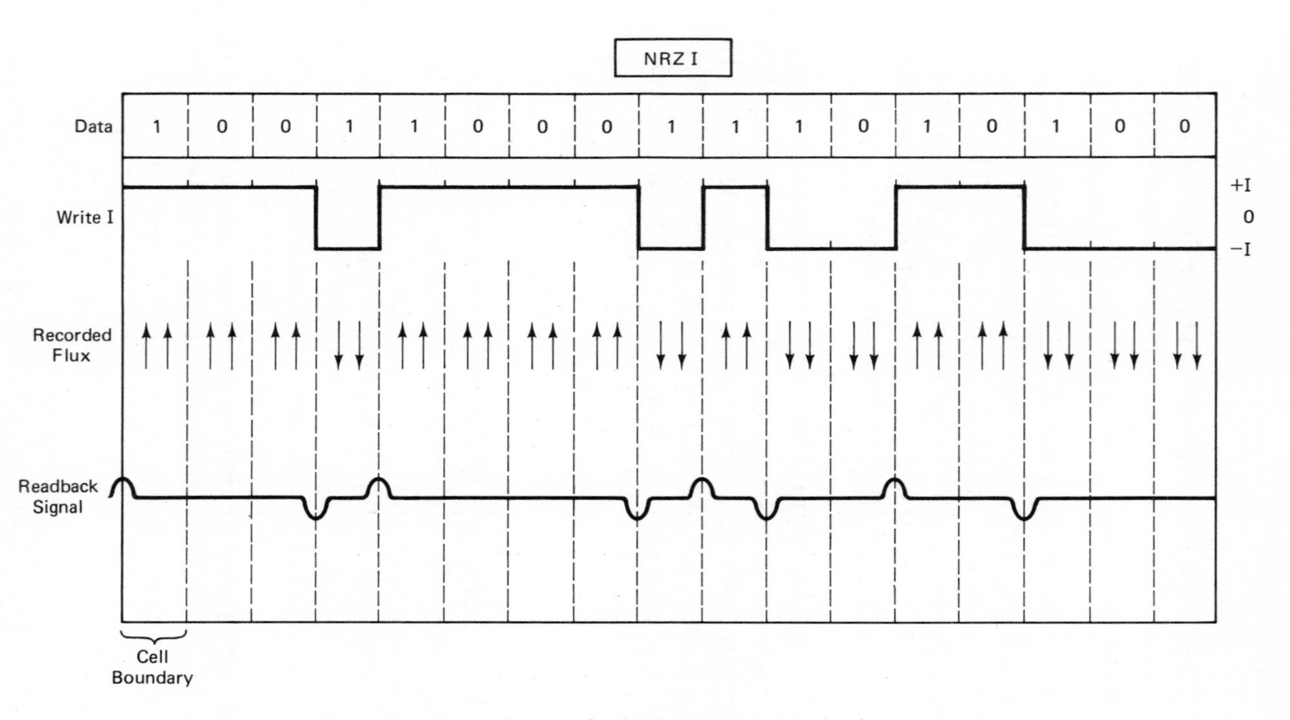

Figure 7-24: A "1" is recorded by a reversal in write current. A "0" is recorded by no reversal of current.

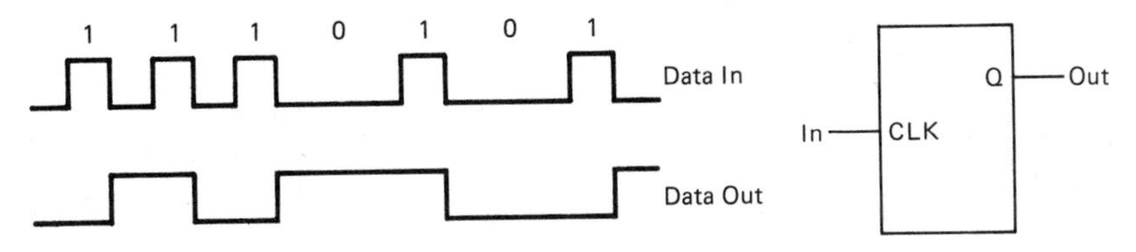

Figure 7-25: NRZI Encoding.

The MFM encoding scheme has the interesting property that an all 1s data pattern and an all 0s data pattern each produces the same flux reversal pattern. When a system uses MFM there must be a method to synchronize the system clock to the data pattern. If this synchronization does not occur, the data cannot be read correctly.

Comparisons

Figure 7-28 shows a binary data pattern and how it would look when encoded into NRZI, FM, and MFM. From this direct comparison, it can be seen the NRZI is the most efficient, FM is the most inefficient, and MFM is a compromise between the two.

Current high-density magnetic tape recording uses a scheme called Group Code Recording (GCR). GCR is not a different method of encoding flux reversals on the magnetic surface. Systems using GCR usually employ NRZI as the basic technique for recording. GCR uses an encoder that converts all four-bit binary numbers into five-bit binary numbers as shown in Figure 7-29. The encoded five-bit numbers have the property that there are no more than two zeros in a row. When recorded as NRZI, this insures that there are no long lengths of tape with no flux reversals. This aids synchronization with the system clock.

Two-of-Seven

Current high-density magnetic disc recording also uses various encoding schemes prior to the actual encoding of the magnetic flux reversals. The purpose of these encoding schemes is to increase the ratio of bits per inch (BPI) to flux changes per inch (FCI).

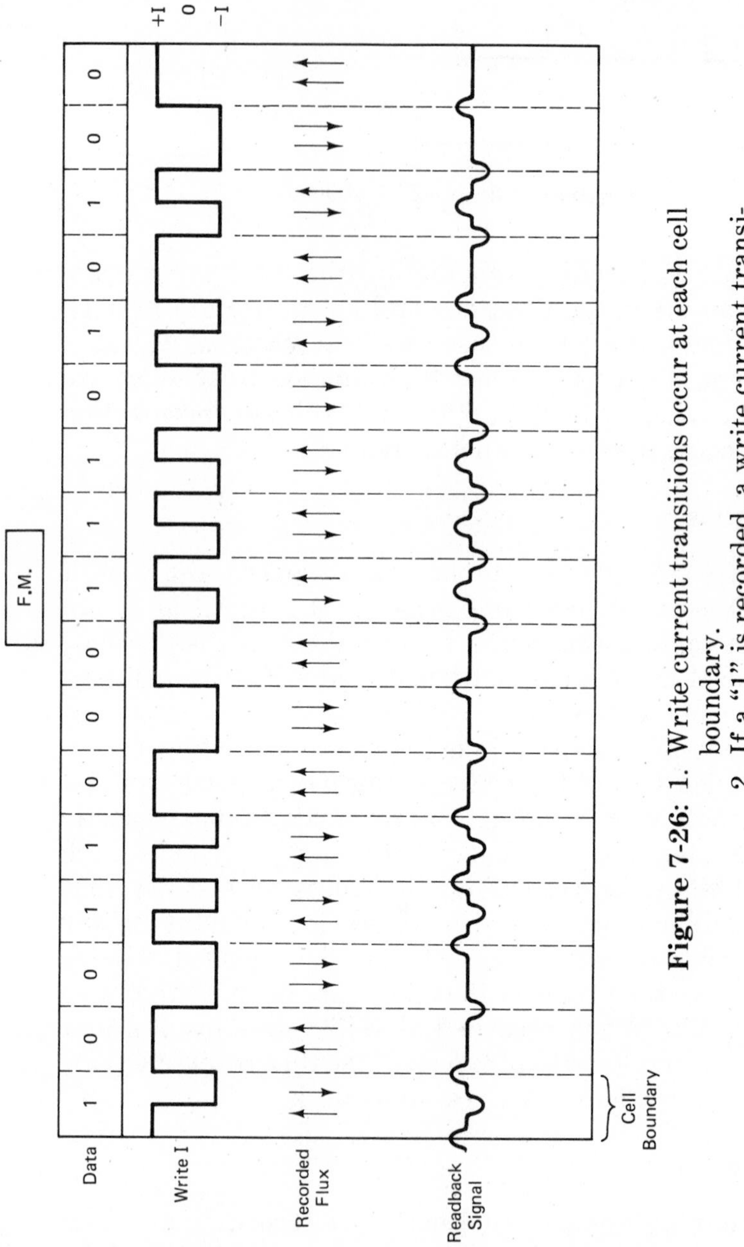

Figure 7-26: 1. Write current transitions occur at each cell boundary.
2. If a "1" is recorded, a write current transition will occur at the half-cell boundary,
3. If a "0" is recorded, no write current transition occurs at the half-cell boundary.
4. If the write current frequency for an all O's pattern is X, the pattern for all 1's is 2X.

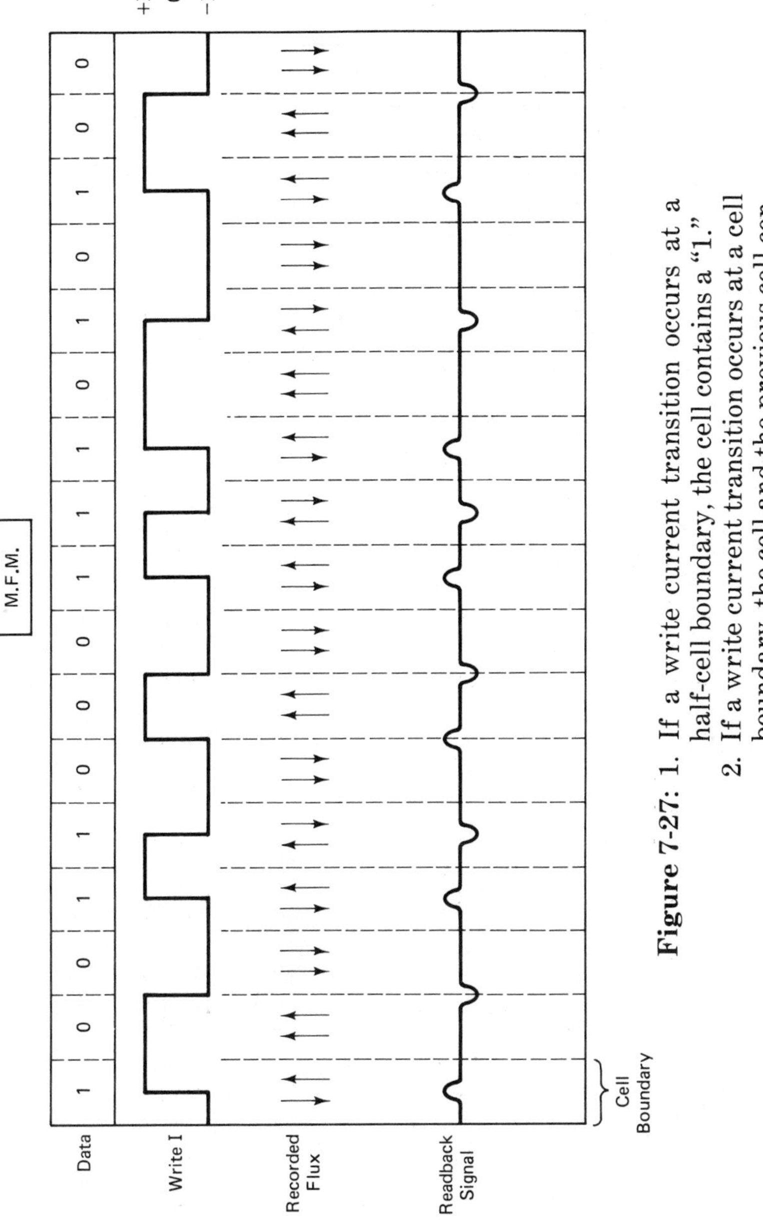

Figure 7-27:
1. If a write current transition occurs at a half-cell boundary, the cell contains a "1."
2. If a write current transition occurs at a cell boundary, the cell and the previous cell contain a "0."
3. If there is no write current transition at the cell boundary or half-cell boundary, the previous cell contains a "0."

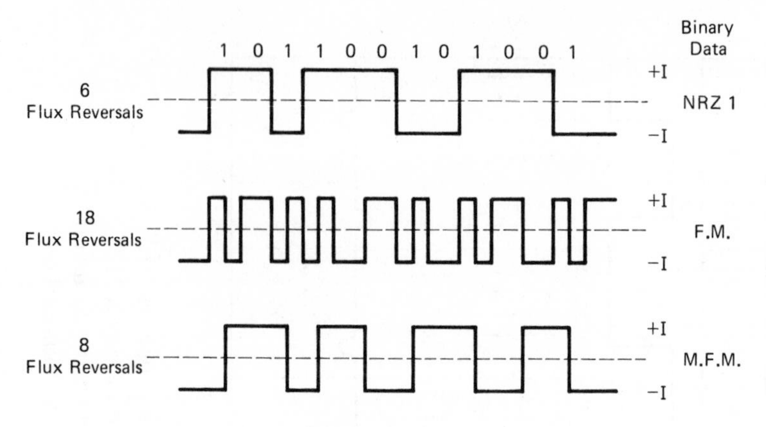

Figure 7-28

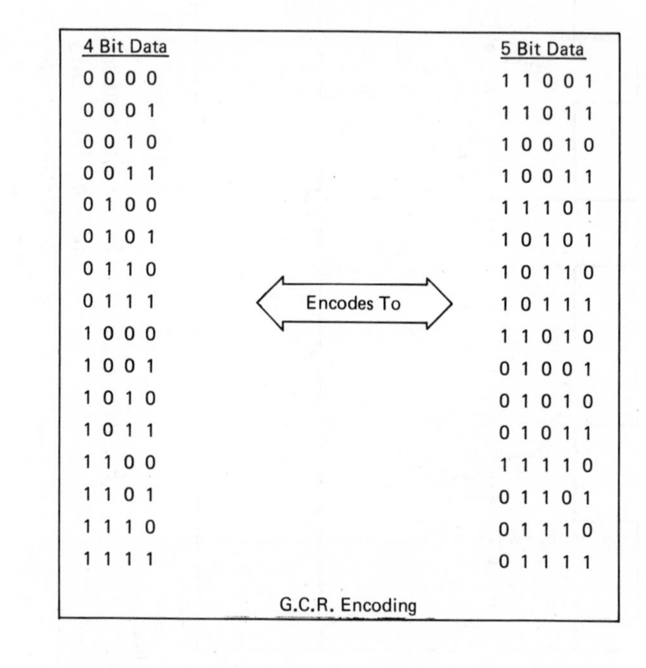

4 Bit Data		5 Bit Data
0 0 0 0		1 1 0 0 1
0 0 0 1		1 1 0 1 1
0 0 1 0		1 0 0 1 0
0 0 1 1		1 0 0 1 1
0 1 0 0		1 1 1 0 1
0 1 0 1		1 0 1 0 1
0 1 1 0		1 0 1 1 0
0 1 1 1	Encodes To	1 0 1 1 1
1 0 0 0		1 1 0 1 0
1 0 0 1		0 1 0 0 1
1 0 1 0		0 1 0 1 0
1 0 1 1		0 1 0 1 1
1 1 0 0		1 1 1 1 0
1 1 0 1		0 1 1 0 1
1 1 1 0		0 1 1 1 0
1 1 1 1		0 1 1 1 1

G.C.R. Encoding

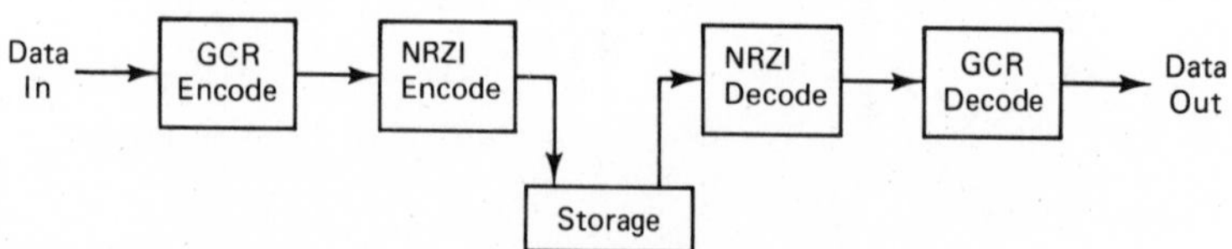

Figure 7-29: G.C.R. Magnetic Recording.

One of the schemes currently in use is called two-of-seven. Two-of-seven encoders have the property that no matter what the input is, the output has at least two zeros between each one. When used in conjunction with NRZI this gives fewer magnetic flux reversals than if the same data were not encoded. Figure 7-30 gives the concept of two-of-seven encoding. The binary data is processed through the encoder and then recorded on the disc as NRZI. Some eight-bit binary numbers are shown along with their decimal equivalents. The output of the encoder is shown for each pattern.

0 0 0 0 0 0 0 0	0	0 0 0 1 0 0 0 0 0 1 0 0 0 0 0 1
1 1 1 1 1 1 1 1	255	1 0 0 0 1 0 0 0 1 0 0 0 1 0 0 0
0 1 0 0 1 0 0 1	73	1 0 0 1 0 0 1 0 0 1 0 0 1 0 0 1
1 0 0 0 1 1 0 0	140	0 1 0 0 0 0 0 0 1 0 0 0 0 0 0 1

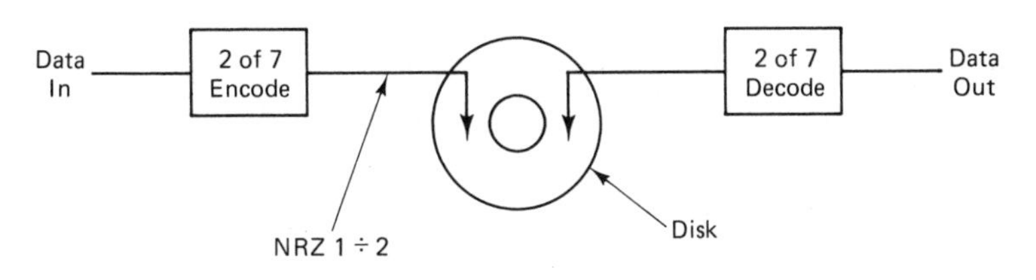

Figure 7-30: The Two-of-Seven Code.

8

How To Interface With the Computer

INTRODUCTION

The computer itself operates at high speeds with binary (or coded binary) numbers. It is capable of performing a vast number of manipulations with binary numbers in a very short time period. The operations that require lots of time, and therefore, limit the speed of the computer are the operations that require interface with the human operators. These operations include data input from terminals, data output on printers, and loading tape and disc memory devices in order to read data.

Most interfacing with the computer requires the use of an alphanumeric code in order to have data in a binary number form. Some of the interfacing requires the use of circuits to convert analog data into digital data and then to convert the manipulated digital data back to analog data. The devices and methods that are used to perform these functions are presented in this chapter.

CHECKLIST OF COMMON ALPHANUMERIC CODES

ASCII

Perhaps the most common code encountered in digital electronics is ASCII or American Standard Code for Information Interchange. This is a 7-bit code in which numbers, letters, and most of the commonly used symbols can be represented in the form of binary numbers. In some case, ASCII is shown as an 8-bit code but the extra bit is just a parity bit* and does not effect the code itself. The ASCII code is defined in Figure 8-1.

5-Bit Baudot Code

The 5-bit Baudot code is named after the inventor Jean Baudot. It is used mainly with teletype devices and data storage devices such as cassette tapes. The 5-bit Baudot code is shown in Figure 8-2. The code can also have a parity bit attached.

The Hollerith Code

The Hollerith code is found on the "punched cards" used for computer input/output. Figure 8-3 shows the Hollerith code for alphanumeric characters as well as various symbols.

The Extended Binary-Coded-Decimal Interchange Code (EBCDIC).

The EBCDIC code is used on large IBM computer systems, as well as other large systems. It is capable of representing numbers, letters, and other special characters. Figure 8-4 shows the EBCDIC code.

Typical Format on Paper Tape

Paper tape devices typically use a five-channel tape or an eight-channel tape. Five-channel paper tape almost always uses the five-bit Baudot code shown in Figure 8-2. Eight-channel paper tape format is shown in Figure 8-5. The position of the parity bit may vary depending on individual systems.

---

*For parity bit explanation, see Chapter 9.

	000	001	010	011	100	101	110	111
0000	NULL	(1) DC_0	$\bar{D}$	0	@	P		
0001	SOM	DC_1	!	1	A	Q		
0010	EOA	DC_2	))	2	B	R		
0011	EOM	DC_3	#	3	C	S		
0100	EOT	DC_4 Stop	$	4	D	T		
0101	WRU	ERR	%	5	E	U		
0110	RU	SYNC	&	6	F	V		
0111	BELL	LEM	)	7	G	W		
1000	FE_0	S_0	(	8	H	X		
1001	HT / SK	S_1	)	9	I	Y		
1010	LF	S_2	*	:	J	Z		
1011	V_{TAB}	S_3	+	;	K	[		
1100	FF	S_4	Comma ,	<	L	\		ACK
1101	CR	S_5	–	=	M	]		(2)
1110	SO	S_6	★	>	N	↑		ESC
1111	SI	S_7	/	?	O	←		DEL

	Top	Side		
Example	100	1000	=	H

Blank Spaces are Unassigned

Abbreviations

NULL	NULL Idle
SOM	Start of Message
EOA	End of Address
EOM	End of Message
EOT	End of Transmission
WRU	Who Are You
RU	Are You ?
BELL	Audible Bell
FE	Format Effector
HT	Horizontal Tabluation
SK	Skip
LF	Line Feed
V_{TAB}	Vertical Tabulation
DC_0	Device Control
FF	Form Feed
CR	Carriabe Return
SO	Shift Out
SI	Shift In
DC_1–DC_3	Device Control
ERR	Error
SYNC	Synchronous Idle
LEM	Logical End of Media
ACK	Acknowledge
(2)	Unassigned Control
ESC	Escape
DEL	Delete Idle
SO_0–SO_7	Separators

Figure 8-1: A.S.C. II Code.

Letters	Figures	Perforated Tape	Binary
A	—	○ ○ ○ · ● ●	0 0 0 1 1
B	?	● ● ○ · ○ ●	1 1 0 0 1
C	:	○ ● ● · ● ○	0 1 1 1 0
D	$	○ ● ○ · ○ ●	0 1 0 0 1
E	3	○ ○ ○ · ○ ●	0 0 0 0 1
F	!	○ ● ● · ○ ●	0 1 1 0 1
G	&	● ● ○ · ● ○	1 1 0 1 0
H	=	● ○ ● · ○ ○	1 0 1 0 0
I	8	○ ○ ● · ● ○	0 0 1 1 0
J	'	○ ● ○ · ● ●	0 1 0 1 1
K	(	○ ● ● · ● ●	0 1 1 1 1
L	)	● ○ ○ · ● ○	1 0 0 1 0
M	,	● ● ● · ○ ○	1 1 1 0 0
N	.	○ ● ● · ○ ○	0 1 1 0 0
O	9	● ● ○ · ○ ○	1 1 0 0 0
P	Ø	● ○ ● · ● ○	1 0 1 1 0
Q	1	● ○ ● · ● ●	1 0 1 1 1
R	4	○ ● ○ · ● ○	0 1 0 1 0
S	BELL	○ ○ ● · ○ ●	0 0 1 0 1
T	5	● ○ ○ · ○ ○	1 0 0 0 0
U	7	○ ○ ● · ● ●	0 0 1 1 1
V	;	● ● ● · ● ○	1 1 1 1 0
W	2	● ○ ○ · ● ●	1 0 0 1 1
X	/	● ● ● · ○ ●	1 1 1 0 1
Y	6	● ○ ● · ○ ●	1 0 1 0 1
Z	))	● ○ ○ · ○ ●	1 0 0 0 1
BLANK		○ ○ ○ · ○ ○	0 0 0 0 0
L.F.		○ ○ ○ · ● ○	0 0 0 1 0
SPACE		○ ○ ● · ○ ○	0 0 1 0 0
C.R.		○ ● ○ · ○ ○	0 1 0 0 0
FIGURES		● ● ○ · ● ●	1 1 0 1 1
LETTERS		● ● ● · ● ●	1 1 1 1 1

Figure 8-2: Five Bit Baudot Code.

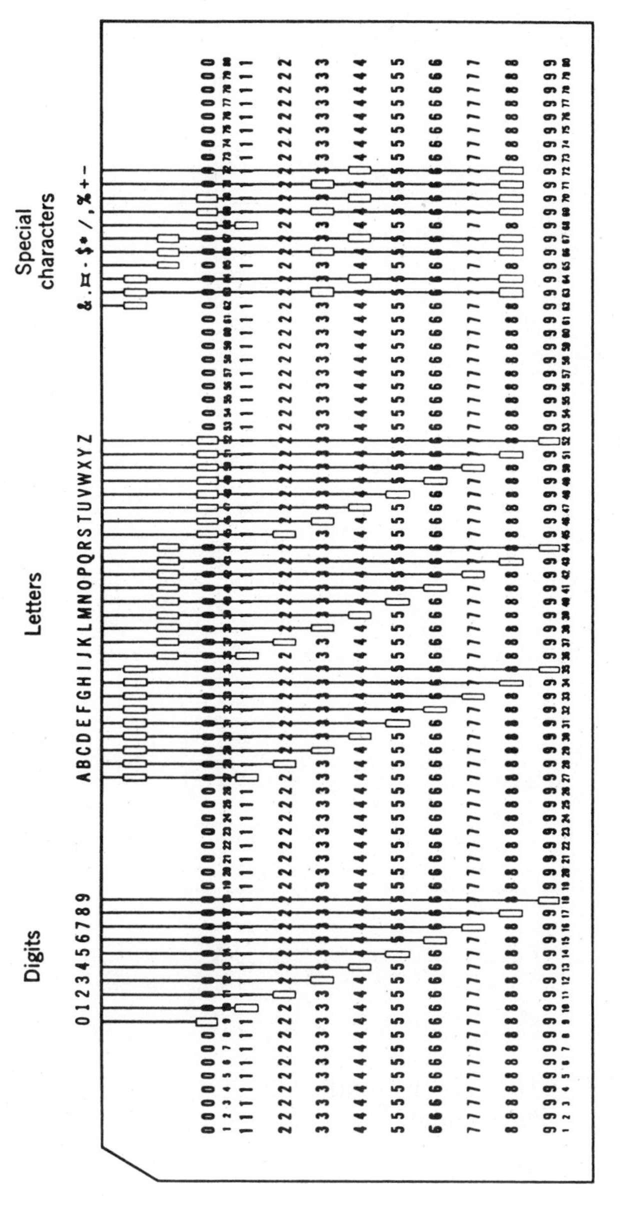

Figure 8-3: Hollerith Code.

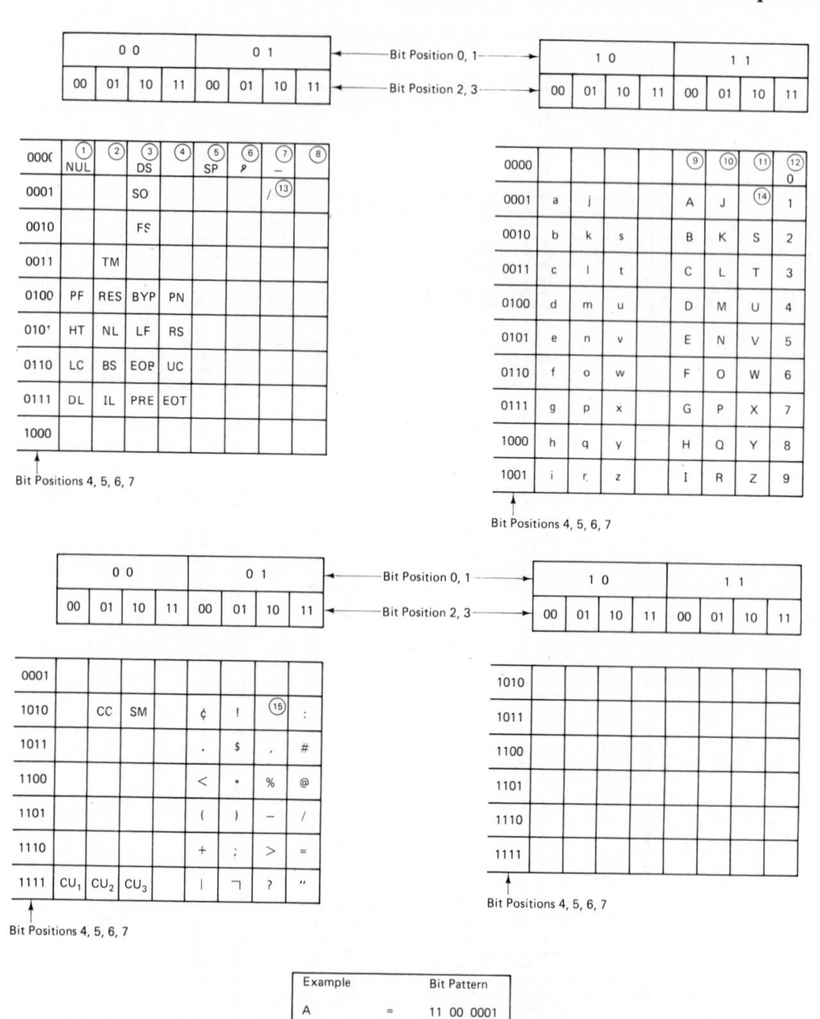

Bit Positions 4, 5, 6, 7	00 / 00	00 / 01	00 / 10	00 / 11	01 / 00	01 / 01	01 / 10	01 / 11
0000	(1) NUL	(2)	(3) DS	(4)	(5) SP	(6) &	(7) –	(8)
0001			SO				/ (13)	
0010			FS					
0011		TM						
0100	PF	RES	BYP	PN				
0101	HT	NL	LF	RS				
0110	LC	BS	EOB	UC				
0111	DL	IL	PRE	EOT				
1000								

Bit Positions 4, 5, 6, 7	10 / 00	10 / 01	10 / 10	10 / 11	11 / 00	11 / 01	11 / 10	11 / 11
0000					(9)	(10)	(11)	(12) 0
0001	a	j			A	J	(14)	1
0010	b	k	s		B	K	S	2
0011	c	l	t		C	L	T	3
0100	d	m	u		D	M	U	4
0101	e	n	v		E	N	V	5
0110	f	o	w		F	O	W	6
0111	g	p	x		G	P	X	7
1000	h	q	y		H	Q	Y	8
1001	i	r	z		I	R	Z	9

Bit Positions 4, 5, 6, 7	00 / 00	00 / 01	00 / 10	00 / 11	01 / 00	01 / 01	01 / 10	01 / 11
0001								
1010		CC	SM		¢	!	(15)	:
1011					.	$	,	#
1100					<	*	%	@
1101					(	)	–	/
1110					+	;	>	=
1111	CU_1	CU_2	CU_3		|	¬	?	"

Bit Positions 4, 5, 6, 7	10 / 00	10 / 01	10 / 10	10 / 11	11 / 00	11 / 01	11 / 10	11 / 11
1010								
1011								
1100								
1101								
1110								
1111								

(Column headings: Bit Position 0, 1 / Bit Position 2, 3)

Example		Bit Pattern
A	=	11 00 0001
¢	=	01 00 1010

Figure 8-4: EBDCID--Extended Binary-Coded-Decimal Interchange Code.

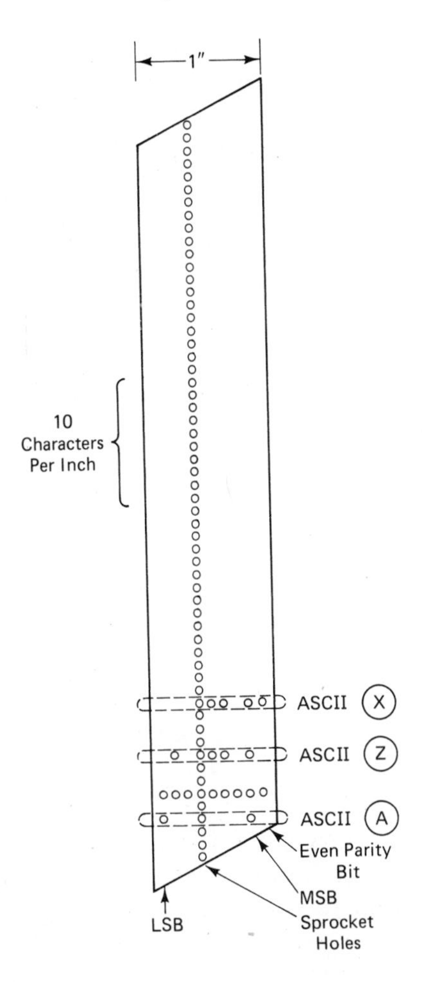

Figure 8-5: Typical 8-Channel Paper Tape Format.

EXAMPLES OF D/A AND A/D CONVERTERS

Digital-to-analog (D/A) and analog-to-digital (A/D) conversion is necessary in order for the computer and other digital devices to deal with the real world. The digital-to-analog conversion is much simpler and can be done with resistor networks. The analog-to-digital conversion is much more complicated and A/D converters usually have a D/A converter as part of their function.

To understand the digital-to-analog converter, consider the resistor network shown in Figure 8-6. The load resistor R_L is made to be much larger than the four input resistors. In that way, the analog output voltage is almost inversely proportional to the value of the input resistor. The input resistors are made "powers of two" lower than R_0 so that the 2, 4, and 8 input will produce that much greater output voltage.

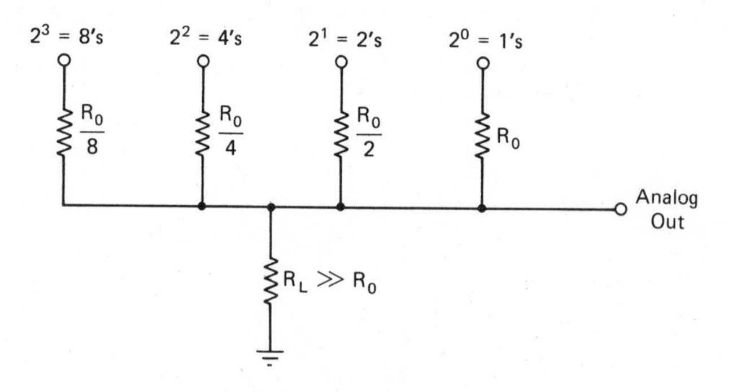

Figure 8-6

The resistor network will produce an output voltage that is almost proportional to the weight of the binary number input. It cannot be exact since a resistor voltage divider has an output:

$$V_{OUT} = \left(\frac{R_L}{R_O + R_L}\right) \times V_{IN}$$

This error can be made small by making R_L much larger than R_O.

This resistor network error can be eliminated by using the op-amp circuit shown in Figure 8-7. The output of this amplifier is:

$$V_{OUT} = -\left(\frac{R_{IN}}{R_L}\right) \times V_{IN}$$

The input resistors are again powers of two lower than R_O to properly weight the inputs.

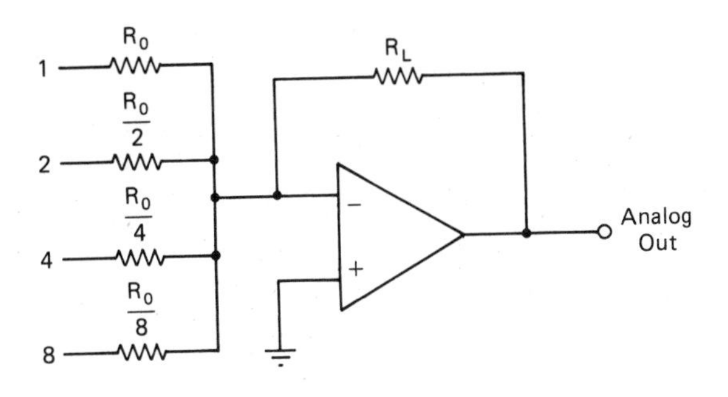

Figure 8-7

To understand the operation of an analog-to-digital converter, consider the circuit shown in Figure 8-8. The analog signal is applied to one input of a voltage comparator. The other input of the comparator comes from a digital-to-analog converter. When the input of the D/A converter is increased, the analog output voltage will increase. When the D/A output comes close to the analog input voltage, the comparator output goes to zero. This will stop the input to the D/A converter and give an appropriate digital output.

The control box is the device which causes the digital input to the D/A converter to increase. This can be done by different means. Figure 8-8(A) shows the increase occurring in regular steps until the D/A output matches the analog input. Figure 8-8(B) shows the control unit taking guesses and converging on the value by comparing each guess to the analog input.

The second method is faster than the first. Systems that require even more speed set up a network of comparators and apply the analog input to each one and then select the proper output. A three-bit simultaneous A/D converter is shown in Figure 8-9. In order to convert to three bits it is necessary to use seven analog comparators.

SOME COMMON INTERFACES

Interfacing for the purpose of transmitting data back and forth is so common that certain standard interfaces have

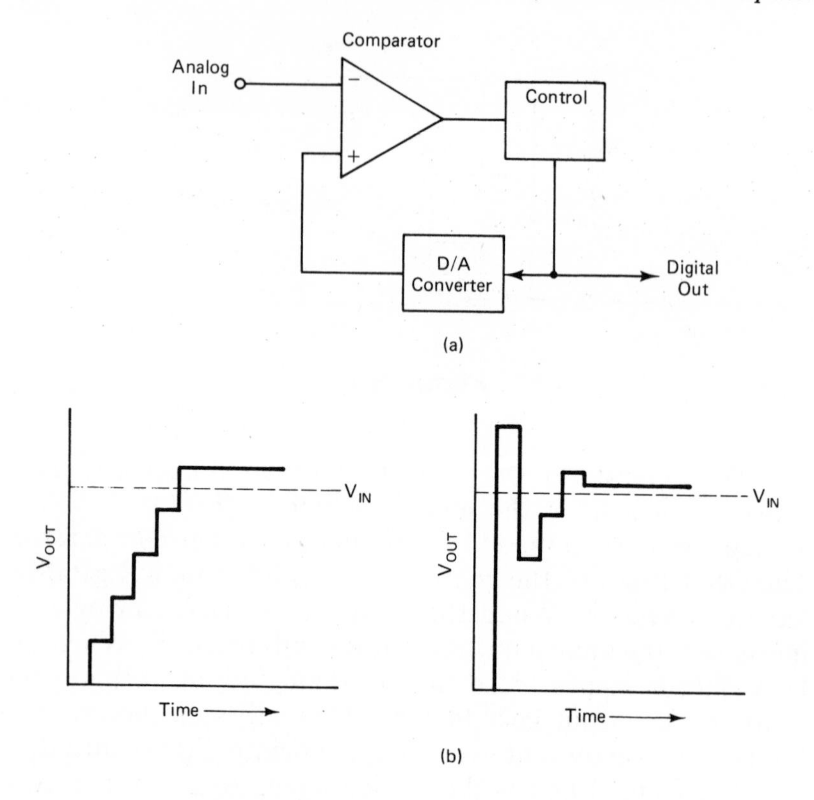

Figure 8-8: A/D Converter.

been adopted. Two of the most common are the RS-232C and the IEEE-488 interfaces. The characteristics of these interfaces are described in the following.

RS-232C

The RS-232C is defined as an Electronic Industries Association (EIA) standard. This standard specifies the interconnection of data terminal equipment (DTE) and data communication equipment (DCE). When used in connection with computers, the computer becomes the DCE, and the peripheral device becomes the DTE.

The interface is very general in its applications. Figure 8-10 shows a typical interconnection between the DCE and

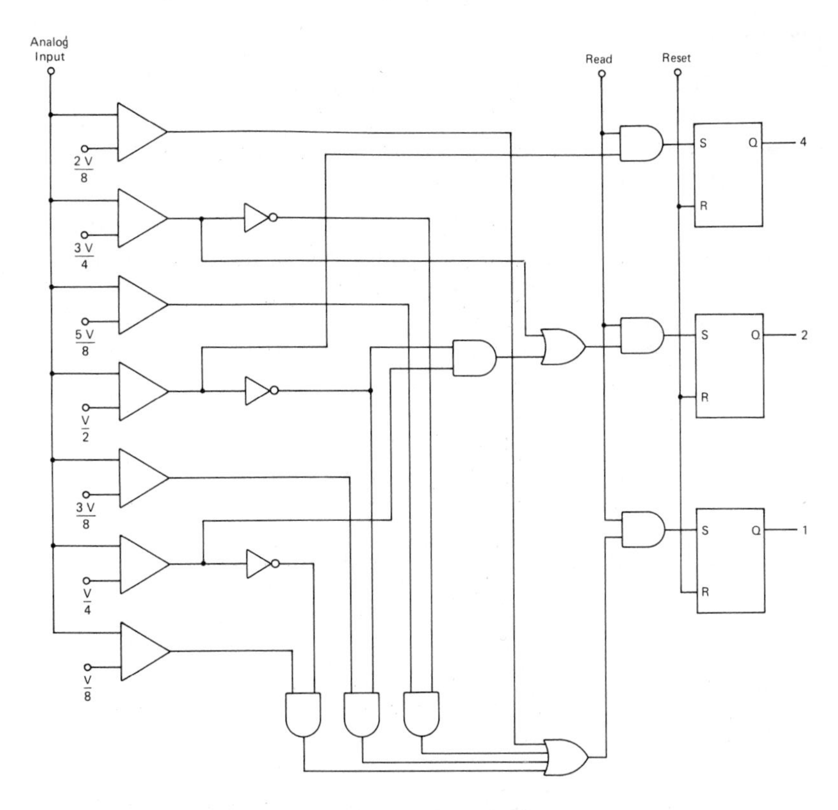

Figure 8-9: Three-Bit Simultaneous A/D Converter.

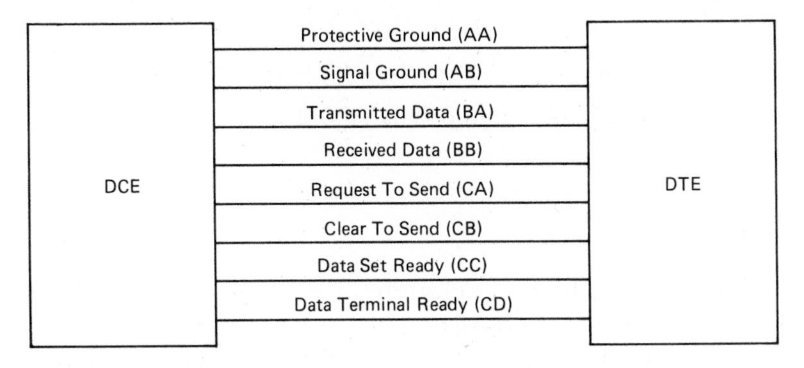

Figure 8-10: Typical RS-232C Communications System.

DTE. The data is transmitted in serial form and there are no restrictions on codes, bit lengths, bit sequence, etc. The cable lengths are intended to be short (50 ft. maximum) and the interface can handle all of the common baud rates up to 20 kilobaud.

There is a maximum of 25 lines on the RS-232C connector, although not all 25 need to be used. A description by pin number is shown in Figure 8-11. The 25 lines are divided into the following:

data signals-------four lines
control signals---twelve lines
timing signals----three lines
grounds ------------two lines
unassigned --------four lines

The logic levels for the data lines and control lines are as follows:

DATA SIGNALS

Binary 1----------−3.0 v to −25.0v
Binary 0----------+3.0v to +25.0v
undefined--------−3.0 v to +3.0 v

CONTROL SIGNALS

on----------−3.0 v to +25.0 v
off---------−3.0 v to −25.0 v

Initialization

There must be an exchange of control signals between DCE and DTE before data can be transmitted. This is called handshaking. Figure 8-12 shows the signals which must occur and the timing relationship between them.

PIN	CIRCUIT	ABBREVIATION	DESCRIPTION
1	AA	—	Protective ground
2	BA	TXD	Transmitted data
3	BB	RXD	Received data
4	CA	RTS	Request to send
5	CB	CTS	Clear to send
6	CC	DSR	Data set ready
7	AB	—	Common return (signal GND)
8	CF	DCD	Received line signal detector
9	—	—	Reserved for data set testing
10	—	—	Reserved for data set testing
11	—	—	Unassigned
12	SCF	—	Secondary received line signal detector
13	SCB	—	Secondary clear to send
14	SBA	—	Secondary transmitted data
15	DB	—	Transmission signal element timing
16	SBB	—	Secondary received data
17	DD	—	Receiver signal element timing
18	—	—	Unassigned
19	SCA	—	Secondary request to send
20	CD	DTR	Data terminal ready

Figure 8-11: RS-232 Pin Assignment.

PIN	CIRCUIT	ABBREVIATION	DESCRIPTION
21	CG	—	Signal quality detector
22	CE	—	Ring detector
23	CH/CI	—	Data signal rate selector
24	DA	—	Transmit signal element timing
25	—	—	Unassigned

Figure 8-11 (continued)

A request to send is raised by the terminal. The DCE must respond with a clear to send, and the terminal must then respond with a data terminal ready and remain high until all the data has been transmitted.

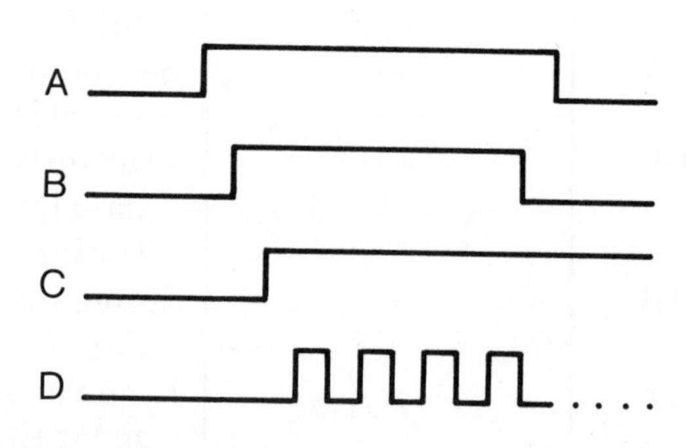

A. Request to SEND from DTE

B. Clear to send from DC

C. Data terminal ready

D. DATA

Figure 8-12

Electrical Characteristics

The equivalent circuit for all RS-232C interfaces is shown in Figure 8-13.

By specification, the following characteristics must apply:

A. R_r must be between 3kΩ and 7kΩ.

B. C_r must be equal to or less than 2.5 nF.

C. The load must not be inductive. This eliminates relays as a load device.

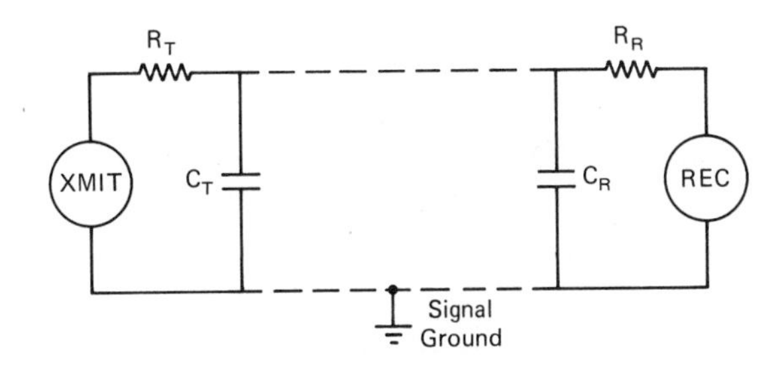

Figure 8-13

The voltage levels used to represent zeros and ones on the RS-232C interface are not the same as those used on standard T^2L logic. Furthermore, to connect the two, there must be an inverter since RS-232C uses negative logic. Typical circuits used to convert RS-232C to T^2L and back are shown in Figure 8-14.

IEEE-488 GENERAL PURPOSE INTERFACE BUS

The IEEE-488 general interface bus is quite different from the RS-232C bus. Its intended use is also quite different although it is very well suited for data transfer between a computer and peripheral equipment.

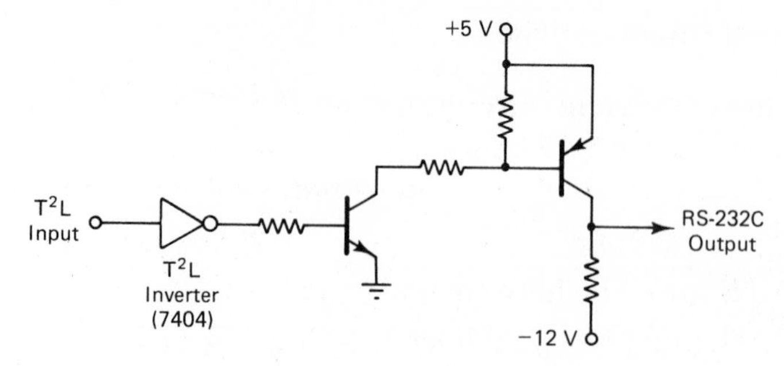

(a) T²L to RS-232C Conversion

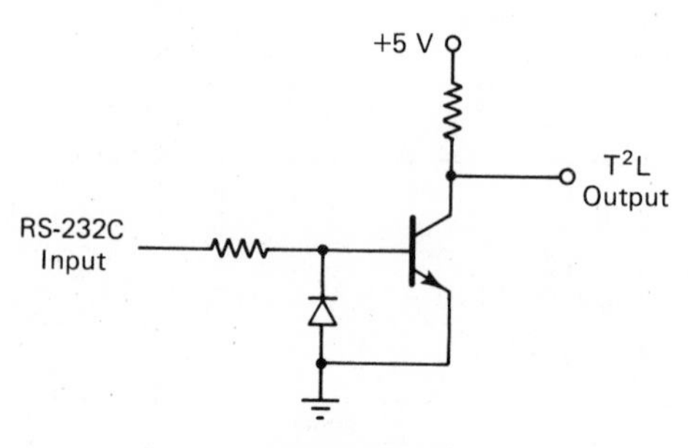

(b) RS-232C to T²L Conversion

Figure 8-14

The general concept of the bus is shown in Figure 8-15. Its characteristics are:

A. It uses parallel data transfer in units of one byte.
B. A maximum of 15 devices can be connected to the bus.
C. The total length of the bus must be less than 60 feet with no more than 6 feet between any two units.
D. The logic levels used on the bus are the same as T²L logic levels.
E. There are 24 pin connections on the line. The pin assignments are shown in Figure 8-16. Eight lines are used for data, three lines are used for handshaking, and five lines are used for general management. The other eight pins are used for return lines (grounds).

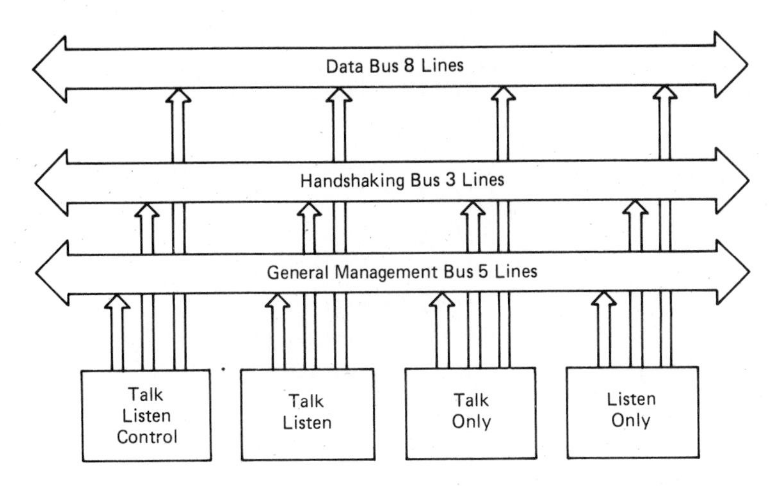

Figure 8-15: IEEE-488 bus Structure.

F. The data and control signals are defined in Figure 8-16(A) and (B).

G. Devices connected to the bus can be talkers (senders) listeners (receivers), controllers or any combination of the three.

Electrical Characteristics

A typical driver/receiver is shown in Figure 8-17. The driver must be capable of sinking 48 MA of current and the Schmitt trigger is used as the receiver in order to enhance noise immunity. All lines are terminated in a resistive load and the capacitive load limit is 100pF within each device. The resistance of each line must not exceed 0.14 ohms/meter and the common logic ground line cannot exceed 0.085 ohms/meter. The shield on the cable is grounded.

THE UART

UART is an acronym for Universal Asynchronous Receiver Transmitter. A simplified block diagram of the device is shown in Figure 8-18. The device is used to convert parallel data into serial data for the purpose of communica-

BUS	LINE	SIGNAL
Data	DIO 1	Data I/O 1
	DIO 2	Data I/O 2
	DIO 3	Data I/O 3
	DIO 4	Data I/O 4
	DIO 5	Data I/O 5
	DIO 6	Data I/O 6
	DIO 7	Data I/O 7
	DIO 8	Data I/O 8
Handshake	DAV	Data Valid
	NDAC	No data accepted
	NRFD	Not ready for data
General management	ATN	Attention
	IFC	Interface clear
	SRQ	Service request
	REN	Remote enable
	EOI	End or identify

(A)

PIN	SIGNAL
1	DIO 1
2	DIO 2
3	DIO 3
4	DIO 4
5	EOI
6	DAV
7	NRFD
8	NDAC
9	IFC
10	SRQ
11	ATN
12	Earth ground (shield)
13	DIO 5
14	DIO 6

(B)

Figure 8-16: Ⓐ IEEE-488 Data; Ⓑ IEEE-488 Pin Connections.

15	DIO 7
16	DIO 8
17	REN
18	DAV (GND)
19	NRFD (GND)
20	NDAC (GND)
21	IFC (GND)
22	SRQ (GND)
23	ATN (GND)
24	LOGIC (GND)

Figure 8-16 (continued)

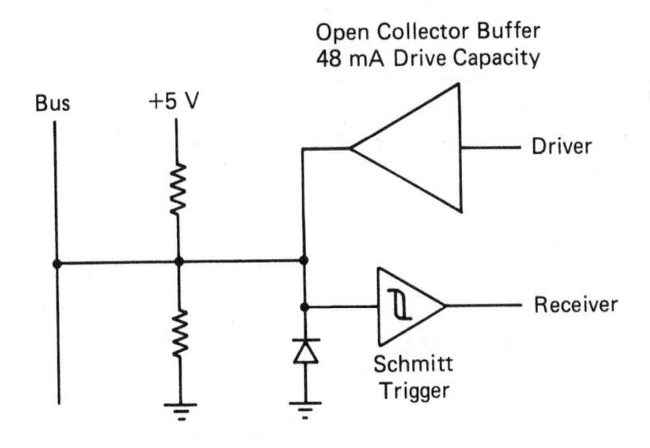

Figure 8-17

tion between computers and peripheral devices. It also controls the baud rate, generates the start and stop bits, and provides other control functions.

As shown in Figure 8-18, the UART consists of a receiver and a transmitter. The receiver and transmitter are connected internally by a bus that allows parallel transfer of data. This bus also connects to the processor. The shift registers in the receiver and transmitter allow for the input and output of data in serial form. The rate of data transfer (baud rate) can be controlled by an external clock.

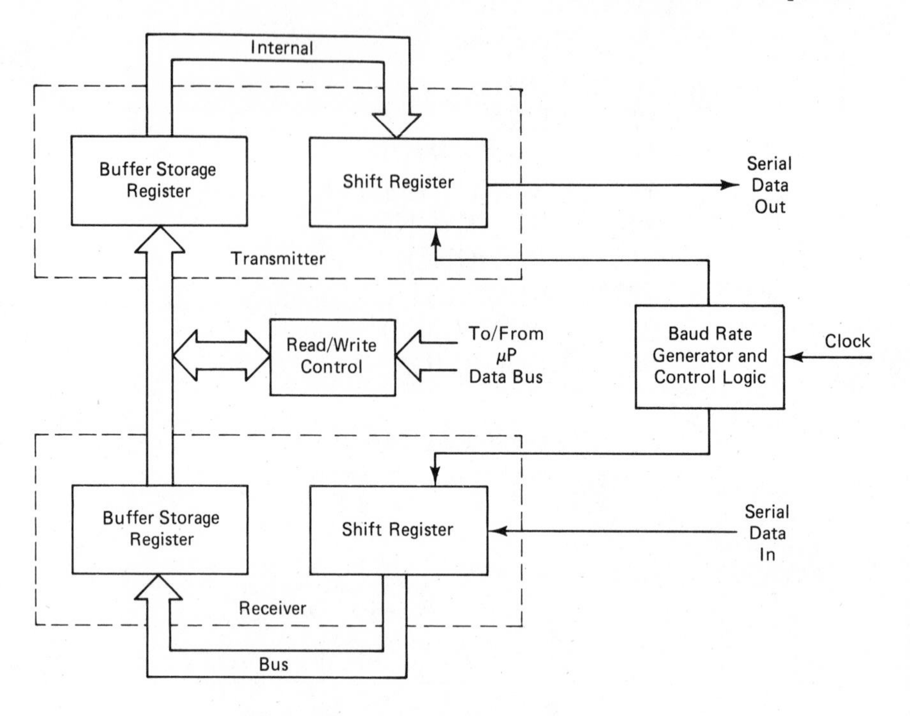

Figure 8-18: Simplified U.A.R.T.

THE TELETYPE INTERFACE

Teletypes are input/output devices consisting of a printer and a keyboard. The printer is an output device and the keyboard can be used to input data to the processor. The teletype is connected to the system through a current loop as shown in Figure 8-19. The current is regulated and usually has a value of 2OMA. This current interface has a very low impedance and is therefore very immune to external noise. 2OMA is used to represent a mark (binary 1) and Øma is used to represent a space (binary 0).

The circuits used to convert T^2L levels to 2OMA current levels on which teletypes operate are shown in Figure 8-20. The optical isolators provide isolation from voltage transients on the line. The Schmitt triggers also provide a degree of noise immunity.

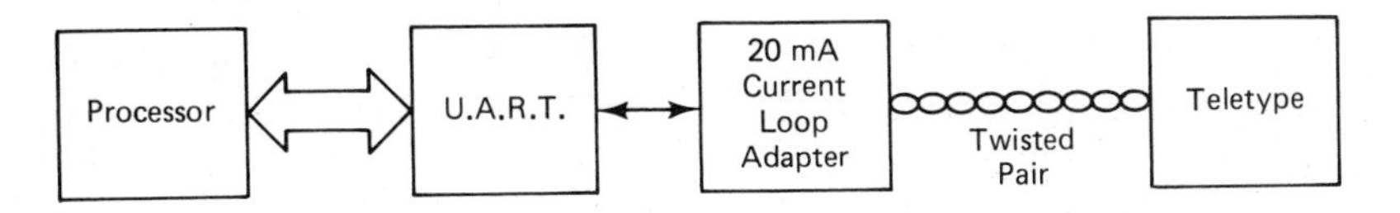

Figure 8-19

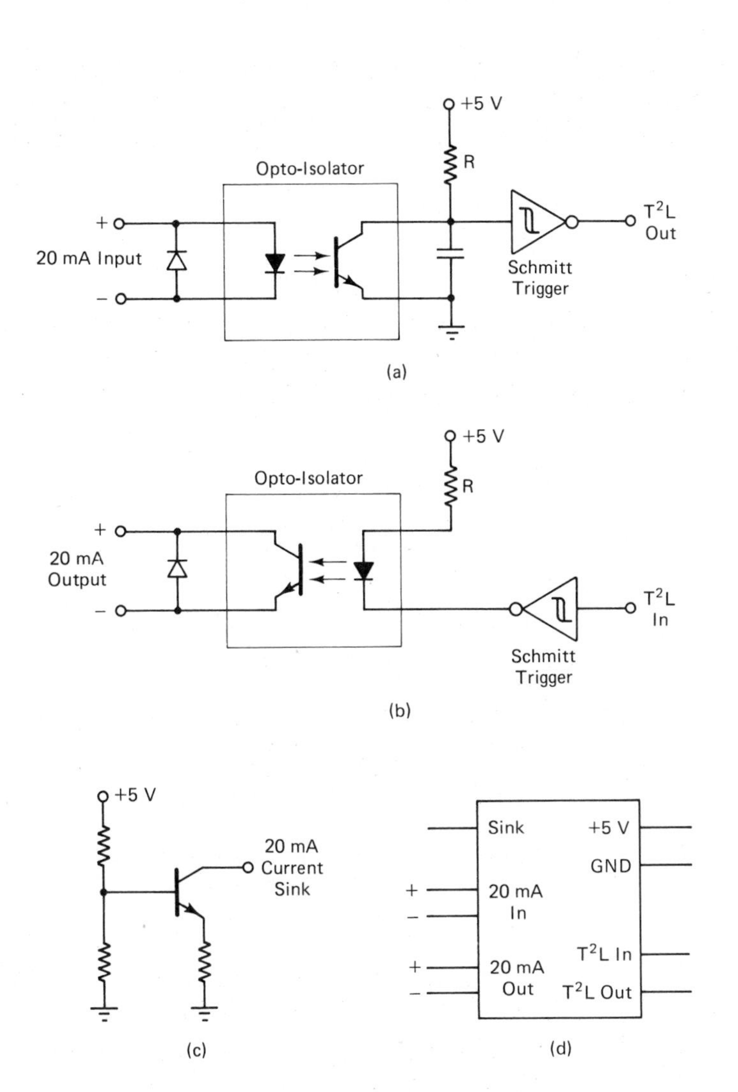

Figure 8-20

A model 33 teletype is a common piece of peripheral equipment. It can send and receive at the rate of ten characters per second. Each character consists of eleven bits. These eleven bits are shown in Figure 8-21. The first bit is a start bit, followed by eight data bits. The last two bits are stop bits.

At a rate of ten characters per second and with eleven bits per character, this gives a data transfer rate of 110 baud. This also produces a time of 9.09 milliseconds for each individual bit. A half-duplex teletype system is shown in Figure 8-22.

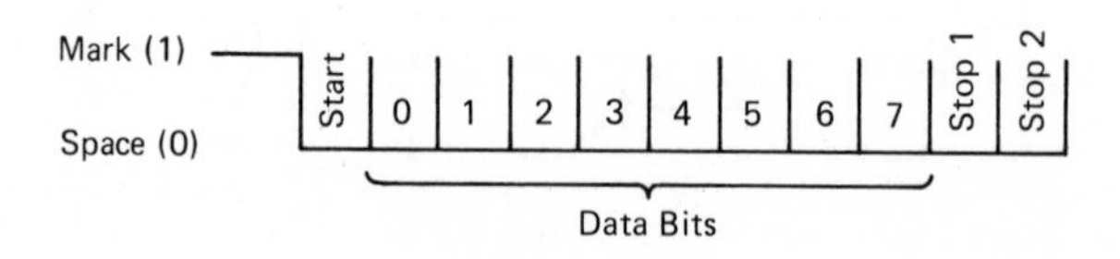

Figure 8-21

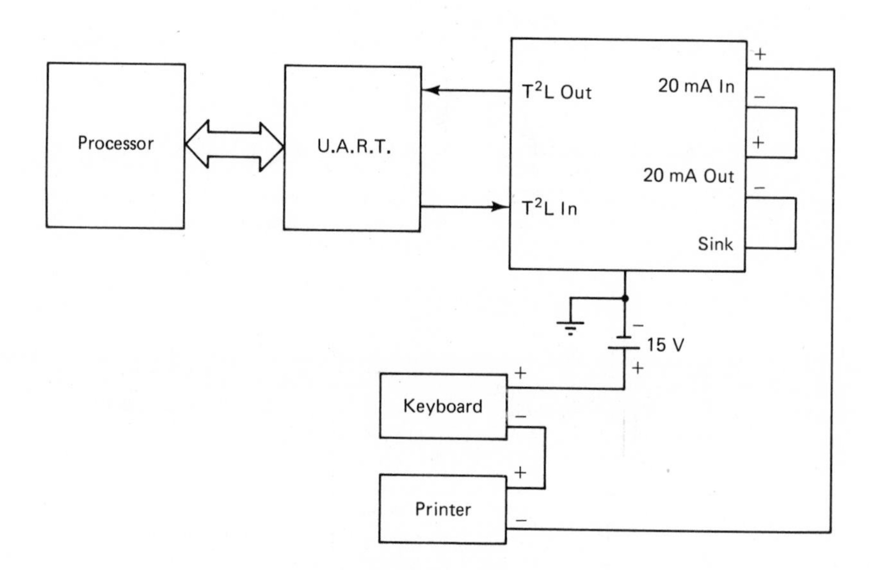

Figure 8-22

9 Understanding and Applying Parity Checks

INTRODUCTION

The process of transmitting data through wires, radio waves, or storing it in memory, is subject to errors. These errors come from a multitude of sources such as static discharges and flaws in memory devices, but the end result is the alteration of some data bits from one to zero and zero to one. All systems are designed to minimize errors, but errors occur anyway. Therefore, it is necessary to devise schemes to detect or even correct these errors before they cause problems.

The parity check scheme of error detection is found throughout the field of digital electronics. It is a relatively simple method of detecting the majority of errors that occur.

DEFINITION OF EVEN/ODD PARITY

The basic concept of parity checks is to add an extra column of bits to the data. This column will contain ones and zeros according to the following rules:

A. For even parity, the parity check column will be used to make the total number of (ones) in that row an even number.

B. For odd parity, the parity check column will be used to make the total number of (ones) in that row an odd number.

For an example of even/odd parity checks, consider the example in Figure 9-1. A set of 4-bit binary data is shown with an extra column for odd parity 9-1(B) and even parity 9-1(C). Note that the only thing necessary to change from even to odd or odd to even parity is an inverter. Either system is equal in its ability to detect errors.

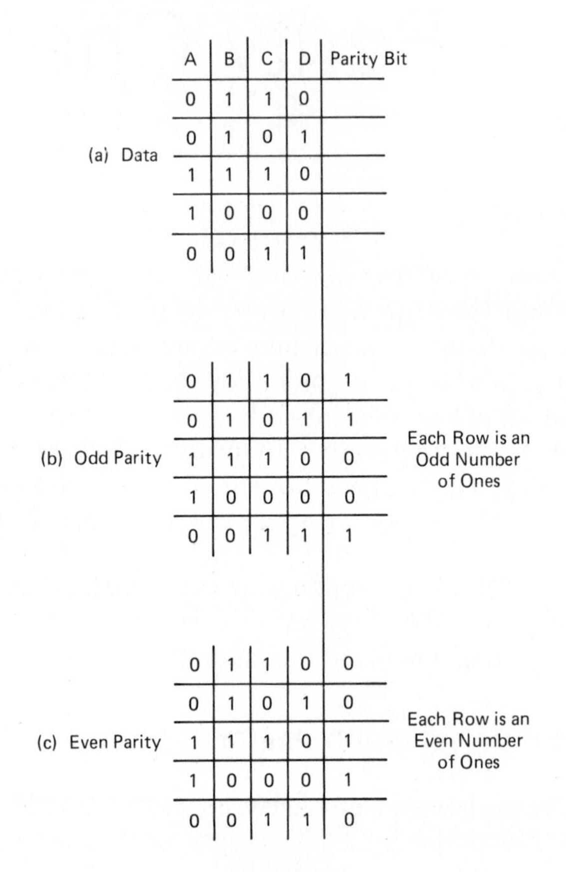

(a) Data

A	B	C	D	Parity Bit
0	1	1	0	
0	1	0	1	
1	1	1	0	
1	0	0	0	
0	0	1	1	

(b) Odd Parity — Each Row is an Odd Number of Ones

A	B	C	D	Parity Bit
0	1	1	0	1
0	1	0	1	1
1	1	1	0	0
1	0	0	0	0
0	0	1	1	1

(c) Even Parity — Each Row is an Even Number of Ones

A	B	C	D	Parity Bit
0	1	1	0	0
0	1	0	1	0
1	1	1	0	1
1	0	0	0	1
0	0	1	1	0

Figure 9-1

EXAMPLES OF PARITY ENCODE/DECODE CIRCUITS

Logic circuits used to generate parity bits and to decode parity bits are made from exclusive OR gates. The exclusive

OR and its truth table are shown in Figure 9-2. The output is the same as modulo 2 addition.

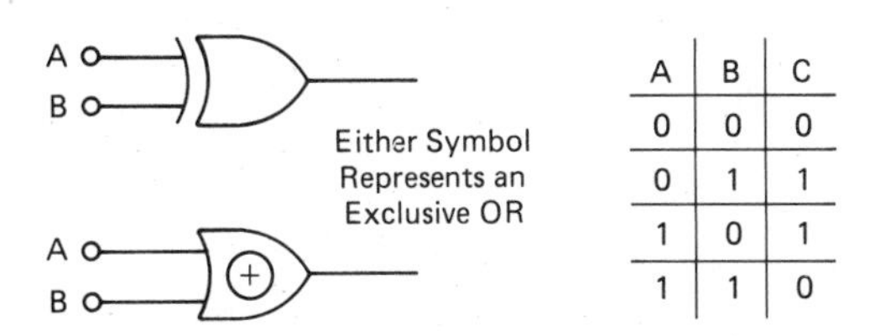

A	B	C
0	0	0
0	1	1
1	0	1
1	1	0

Figure 9-2

For a set of data consisting of two data bits and one parity bit, the transmit and receive circuit would look like Figure 9-3. This example uses even parity. The circuit in Figure 9-4 also uses even parity. It is the same type of circuit as Figure 9-3 except that the data field has been expanded to four bits. The parity check circuit on the receiving end contains the same circuit that generated the parity bit at the transmitter. The parity check circuit operates by regenerating the parity bit from the received data and then compares it to the transmitted parity bit. This regenerate and compare is common to most all error detection and correction schemes.

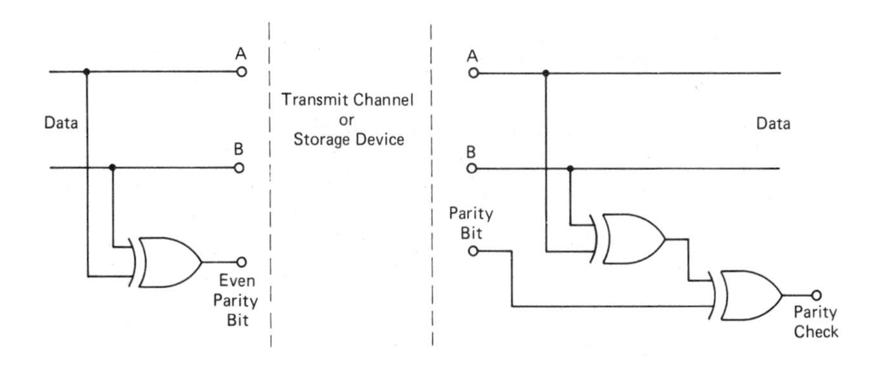

Figure 9-3: If an error occurs, the parity check will be high (binary 1). This error could be either in the data or in the parity bit itself.

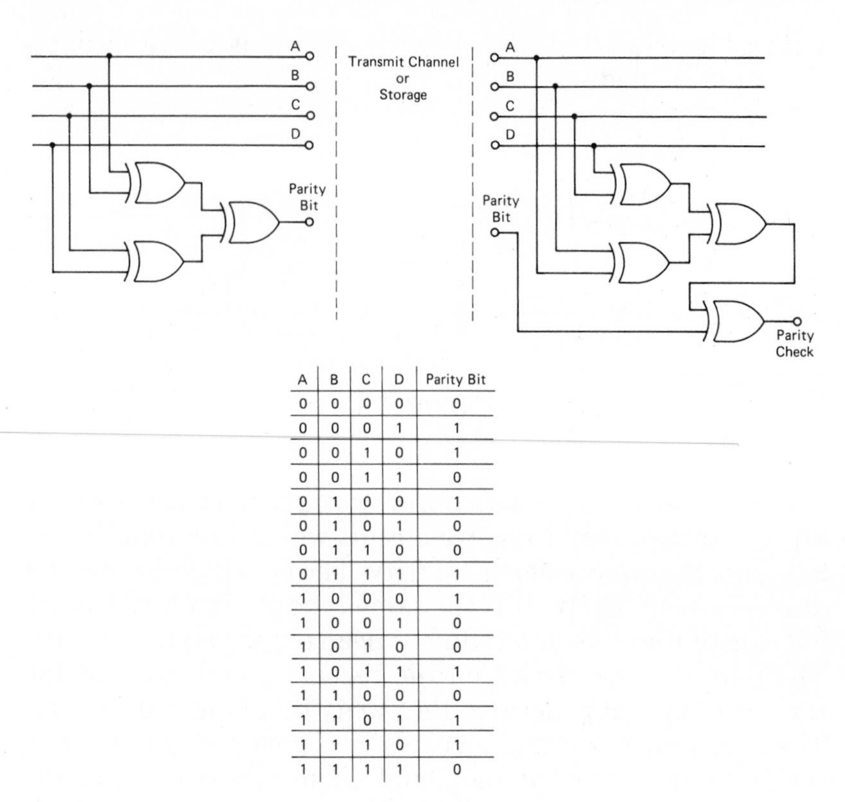

A	B	C	D	Parity Bit
0	0	0	0	0
0	0	0	1	1
0	0	1	0	1
0	0	1	1	0
0	1	0	0	1
0	1	0	1	0
0	1	1	0	0
0	1	1	1	1
1	0	0	0	1
1	0	0	1	0
1	0	1	0	0
1	0	1	1	1
1	1	0	0	0
1	1	0	1	1
1	1	1	0	1
1	1	1	1	0

Figure 9-4: 4-Bit Data with Even Parity Generation and Detection.

LIMITATIONS OF PARITY CHECKS

Parity checks are for error detection only. They cannot correct the error. The normal thing to do when a parity check occurs is to re-transmit the data. If the error cause was transient, the probability is that it will not occur during the second transmission.

Parity checks work only if the error consists of an odd number of bits, i.e., 1, 3, 5, etc. As seen in Figure 9-5, if a two-bit error occurs, the parity check will never know.

Fortunately, in most systems, the probability of multiple bit errors is much smaller than the probability of a single bit error.

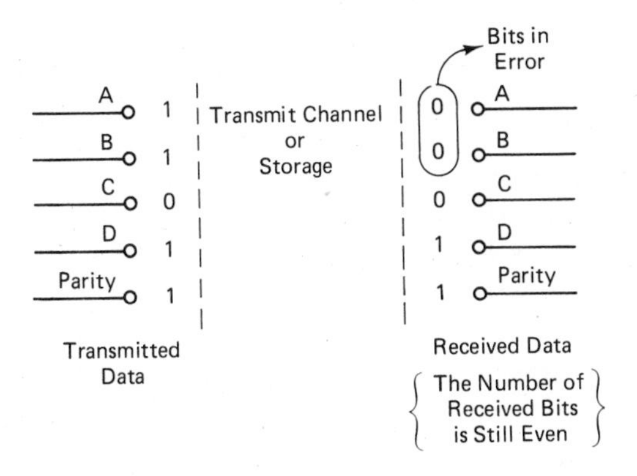

Figure 9-5

THE 74180 PARITY CHECKER

Since parity checks are so common, integrated circuits have been developed to generate and detect parity bits. As seen in circuits shown earlier in this chapter, the parity detect circuit is the same as the generate circuit with the exception of an extra input on the detection end.

The 74180 is capable of generating and detecting either even or odd parity on an eight-bit data word. Its operation is shown in Figure 9-6.

USING MORE COMPLEX PARITY CHECKS

The type of errors that normal parity checks cannot detect usually have a very low probability of occurring. There are times, however, when detection schemes are needed that are more powerful than straight parity. To gain added error detection, it is necessary to increase the number of bits added to the data bits.

Consider the five binary messages in Figure 9-7. If a one-bit error occurs, the parity check should detect it. If a two-bit error occurs in any of the words, the parity check will not detect an error. This nondetectable condition can be overcome by adding a parity check row at the end of the data. The data would now look like Figure 9-8. Any one-bit error would be detected by both the parity column and the parity

Inputs			Output	
Sum	Odd	Even	Odd	Even
Even	0	1	0	1
Odd	0	1	1	0
Even	1	0	1	0
Odd	1	0	0	1

V_{CC} 14 13 12 11 10 9 8 — Data Inputs (13–8)

1 2 — Data Inputs; 3 Even Input; 4 Odd Input; 5 Even Output; 6 Odd Output; 7 GND

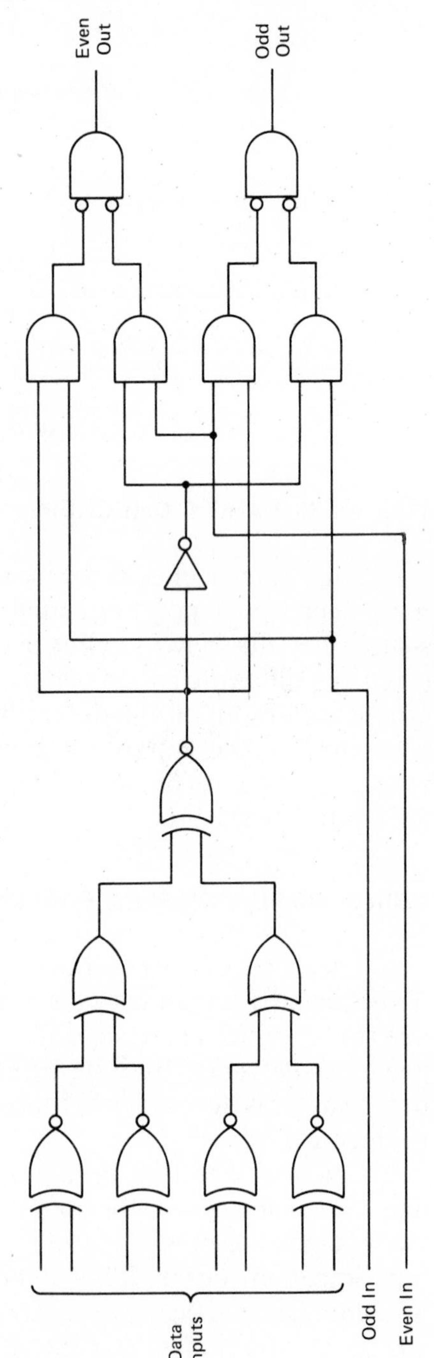

Figure 9-6

				Even Parity
#1	0	1	1	0
2	1	1	0	0
3	0	1	0	1
4	1	0	1	0
5	1	1	1	1

Figure 9-7

#1	0	1	1	0	Parity
2	1	1	0	0	Column
3	0	1	0	1	
4	1	0	1	0	
5	1	1	1	1	
Parity Row →	1	0	1		

Figure 9-8

row. Any two-bit error would be detected by either the parity row or by the parity column, depending on whether the error occurred in a row or in a column.

This parity scheme can detect most errors that occur. However, there are some combinations of errors that cannot be detected. Assume that the first two bits in word 1 and in word 4 are in error, Figure 9-9. As can be seen from the figure, this combination of errors cannot be detected by this method. Another scheme with even more error detection capability involves counting the number of ones in a column or row. The number of ones is then transmitted along with the data shown in Figure 9-10.

The (even) parity bit is added to each row. Below each column is a binary number. This number represents the number of ones in that column. This method detects almost all error combinations that can happen. It is not absolute, because if bits 2 and 3 of word 1 and bits 2 and 3 of word 3 were all in error, this method would not detect an error.

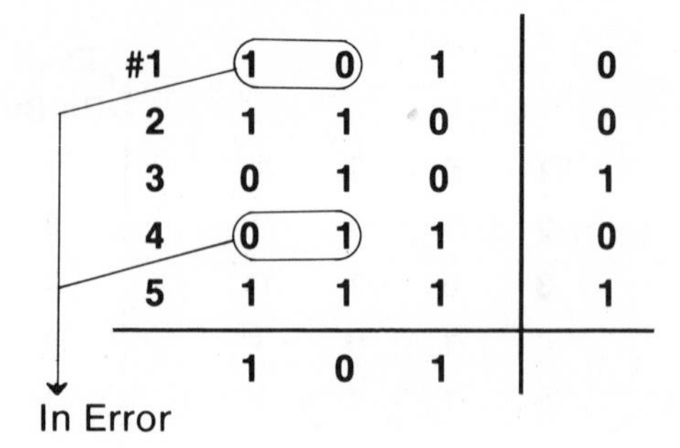

Figure 9-9

					Parity Bit
Word #1	1	0	1	1	1
2	0	0	1	0	1
3	1	1	0	1	1
4	1	0	0	0	1
5	1	0	0	1	0
1's	0	1	0	1	
2's	0	0	1	1	
4's	1	0	0	0	

Figure 9-10

EXAMPLES OF OTHER ERROR CORRECTING SCHEMES

Another method of error detection in common use is known as the 2-out-of-5 code. This method consists of encoding the data according to Figure 9-11 before transmission.

Each encoded binary word consists of five bits with three zeros and two ones. The error detection ability of this code comes from the fact that there are only two ones in each word. If the received word contains something besides two ones, the message is in error. The encoded words also have even parity and a parity circuit can be used to detect an error.

To encode large numbers into 2-out-of-5, encode each separately as shown in the following example.

Example: Encode 1349_{10}

1	3	4	9	Base 10
00101	01001	01010	11000	2-out-of-5

Decimal	Binary	2-out-of-5
0	0 0 0 0	0 0 0 1 1
1	0 0 0 1	0 0 1 0 1
2	0 0 1 0	0 0 1 1 0
3	0 0 1 1	0 1 0 0 1
4	0 1 0 0	0 1 0 1 0
5	0 1 0 1	0 1 1 0 0
6	0 1 1 0	1 0 0 0 1
7	0 1 1 1	1 0 0 1 0
8	1 0 0 0	1 0 1 0 0
9	1 0 0 1	1 1 0 0 0

Figure 9-11

10 Using Hamming Codes

INTRODUCTION

Parity checks are very common in digital electronics. They require only one extra bit to be added to each data block and they detect almost all of the errors that occur. The difficulty with parity is that error correction requires retransmission of the data. An encoding scheme that is able to detect and correct errors, without retries, would have an obvious advantage even if it required more than one check bit to be added to the data.

Hamming codes have the ability to detect and correct errors, and they can be implemented with simple "off-the-shelf" logic chips. This chapter defines these codes, shows how they work, and shows the logic circuits used to implement them.

DEFINITION OF HAMMING CODES

Hamming codes are a subset of ECC codes known as nearest neighbor codes. Data is transmitted using a known number of words, if 4 bits are used there will be 16 possible words; if 8 bits then 256 words, etc. If the received data is not one of the possible words, then an error has occurred and the incorrect data will be corrected by changing it to the data word it most closely resembles, i.e., "nearest neighbor."

Consider the data shown in Figure 10-1, there are three 5-bit words. If these words are compared (bit-by-bit) it is found that words 1 and 2 differ in three places, 1 & 3 in four places, and 2 & 3 in three places. Using ECC terminology, this set of data is said to have a minimum distance of three since no two words differ by less than three bits. If a one-bit error occurs in any of these words, it will be only one bit removed from the correct word while at the same time it will be at least two bits removed from the other words. It can, therefore, be corrected to the nearest neighbor.

	A	B	C	D	E
Word #1	1	0	1	0	1
Word #2	1	1	0	1	1
Word #3	0	1	1	1	0

Figure 10-1

If the received word was 10111, it would be detected as an error and corrected by being changed to word #1.

The illustrations in Figure 10-2 demonstrate nearest neighbor decoding. A and B represent words that are separated by 3 bits. The circles in Figure 10-2(A) represent all possible words removed by one bit from A or B. Since these circles do not overlap, all possible one-bit errors can be corrected. The circles in Figure 10-2(B) represent all possible words removed by two bits from A or B. The area where the two circles overlap represent errors that cannot be corrected since there is no way to tell whether the word should have been A or B.

The minimum distance between data words is the sole determinant as to how many bits can be corrected. This relationship is shown in Figure 10-3.

Hamming codes consist of check bits added to the end of each word. The words themself remain intact. The check bits are chosen so that the minimum distance between words

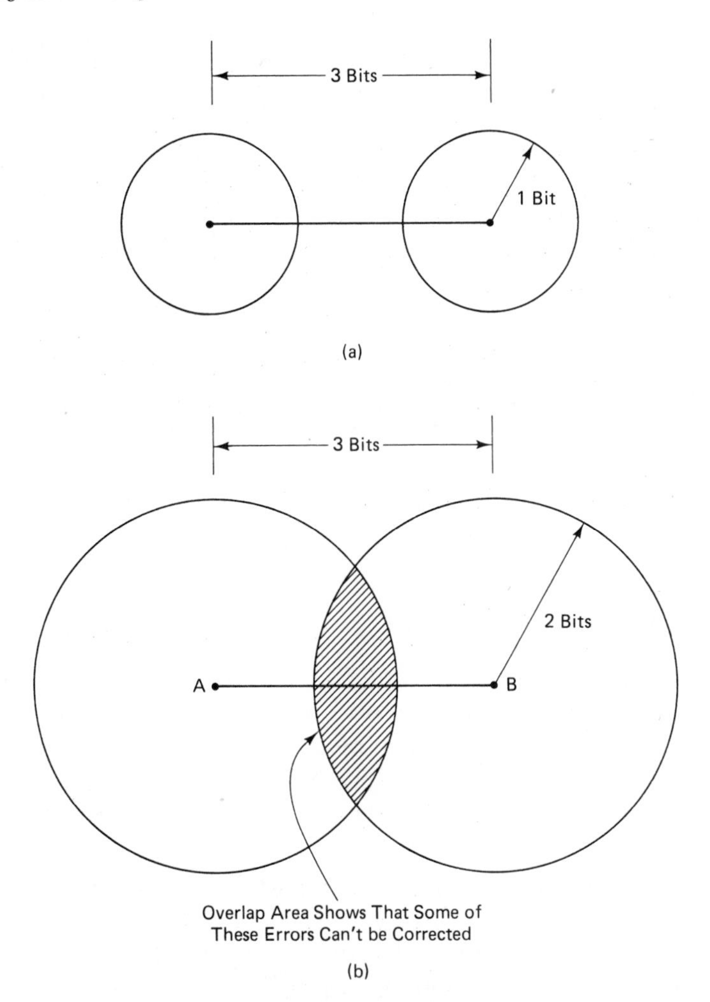

Figure 10-2

Minimum Distance	Bits Correctable
3 To 4	1
5 To 6	2
7 To 8	3
9 To 10	4
11 To 12	5

Figure 10-3

(after the check bits are added) is three. This makes the Hamming codes single-bit error correcting. The following are definitions of the terms commonly used.

Hamming weight—the number of 1's in the code word.

Hamming distance—the number of bits that are different between two code words.

Minimum distance—the minimum Hamming distance between any two code words in a set of data.

(N,K) code—K is the number of data bits. N is the total number of bits. Therefore, N-K is the number of check bits.

Syndrome—Local check bits + regenerated check bits. A non-zero syndrome indicates an error.

THE (7, 4) HAMMING CODES

The (7, 4) Hamming code consists of adding three check bits to a four-bit word. The total word length becomes seven bits. The minimum distance of the (7, 4) Hamming code is three, which means that this code is able to detect and correct all one-bit errors. It can also detect all two-bit errors.

Messages are sent with 3 check bits added to each 4 bit word.

The code is shown in Figure 10-4 with the three check bits that correspond to each four-bit word. It is possible to compare each possible combination of messages to see what the Hamming distance is. This is done in Figure 10-5. It is clear from this comparison that the minimum distance is three.

The set of words in Figure 10-4 represents the (7, 4) Hamming code to be used throughout this chapter. It is not the only manner in which the three check bits can be arranged and still produce a minimum distance of three. Another possibility is shown in Figure 10-6. Any manner in which the check bits are arranged that produce a minimum distance of three are known as (7, 4) Hamming codes.

THE 7, 4 HAMMING CODE

	16-4 bit words	3 check bits added to each word
A	0 0 0 0	0 0 0
B	0 0 0 1	0 1 1
C	0 0 1 0	1 1 0
D	0 0 1 1	1 0 1
E	0 1 0 0	1 1 1
F	0 1 0 1	1 0 0
G	0 1 1 0	0 0 1
H	0 1 1 1	0 1 0
I	1 0 0 0	1 0 1
J	1 0 0 1	1 1 0
K	1 0 1 0	0 1 1
L	1 0 1 1	0 0 0
M	1 1 0 0	0 1 0
N	1 1 0 1	0 0 1
O	1 1 1 0	1 0 0
P	1 1 1 1	1 1 1

Figure 10-4

Min Dist	Min Dist	Min Dist
AB-3	DE-4	HI-7
AC-3	DF-3	HJ-4
AD-4	DG-3	HK-4
AE-4	DH-4	HL-3
AF-3	DI-3	HM-3
AG-3	DJ-4	HN-4
AH-4	DK-4	HO-4
AI-3	DL-3	HP-3
AJ-4	DM-7	IJ-3
AK-4	DN-4	IK-3
AL-3	DO-4	IL-4
AM-3	DP-3	IM-4
AN-4	EF-3	IN-3
AO-4	EG-3	IO-3
AP-7	EH-4	IP-4
BC-4	EI-3	JK-4
BD-3	EJ-4	JL-3
BE-3	EK-4	JM-3
BF-4	EL-7	JN-4
BG-4	EM-3	JO-4
BH-3	EN-4	JP-3
BI-4	EO-4	KL-3
BJ-3	EP-3	KM-3
BK-3	FG-4	KN-4
BL-4	FH-3	KO-4
BM-4	FI-4	KP-3
BN-3	FJ-3	LM-4
BO-7	FK-7	LN-3
BP-4	FL-4	LO-3
CD-3	FM-4	LP-4
CE-3	FN-3	MN-3
CF-4	FO-3	MO-3
CG-4	FP-4	MP-4
CH-3	GH-3	NO-4
CI-4	GI-4	NP-3

Figure 10-5

Min Dist	Min Dist	Min Dist
CJ-3	GJ-7	OP-3
CK-3	GK-3	
CL-4	GL-4	
CM-4	GM-4	
CN-7	GN-3	
CO-3	GO-3	
CP-4	GP-4	

Figure 10-5 (continued)

	Data Bits	Check Bits
A	0 0 0 0	0 0 0
B	0 0 0 1	1 1 1
C	0 0 1 0	0 1 1
D	0 0 1 1	1 0 0
E	0 1 0 0	1 0 1
F	0 1 0 1	0 1 0
G	0 1 1 0	1 1 0
H	0 1 1 1	0 0 1
I	1 0 0 0	1 1 0
J	1 0 0 1	0 0 1
K	1 0 1 0	1 0 1
L	1 0 1 1	0 1 0
M	1 1 0 0	0 1 1
N	1 1 0 1	1 0 0
O	1 1 1 0	0 0 0
P	1 1 1 1	1 1 1

Figure 10-6

A check of all messages used in the (7, 4) Hamming code shows a minimum distance of three.

EXAMPLE OF A COMPLETE ENCODE/DECODE SYSTEM

The generation and detection of Hamming codes is very similar to the generation and detection of parity checks. Exclusive OR gates are used and, like parity, the code is regenerated on the receiving end and compared to the transmitted code. A complete system for the code given in Figure 10-4 is shown in Figure 10-7. (Also see Figure 10-8 for diagram.)

A Step-by-Step Error Correction

To help understand the functioning of the (7, 4) Hamming code, the following example assumes an error in the received data and follows the signals throughout the circuit.

1. The 4-bit message $X_1, X_2, X_3, X_4 = 0110$ is transmitted.
2. The check bits would be $Y_1, Y_2, Y_3 = 001$ (from Figure 10-4).
3. Assuming a received error in X_4, the received data would be $X_1, X_2, X_3, X_4, Y_1, Y_2, Y_3 = 0111001$.
4. With the error, the re-encoded check bits would be $Y_1, Y_2, Y_3 = 010$.
5. When this is compared to the received check bits, there will be 1 outputs from OR_2 and OR_3.
6. Since the syndrome is not zero there will be an error flag. The output of OR_8 be 1.
7. With the outputs of OR_2 and OR_3 high, the 3:8 decoder will have the following outputs:

Pins	Output
5	1
7	1
6	1
3	0

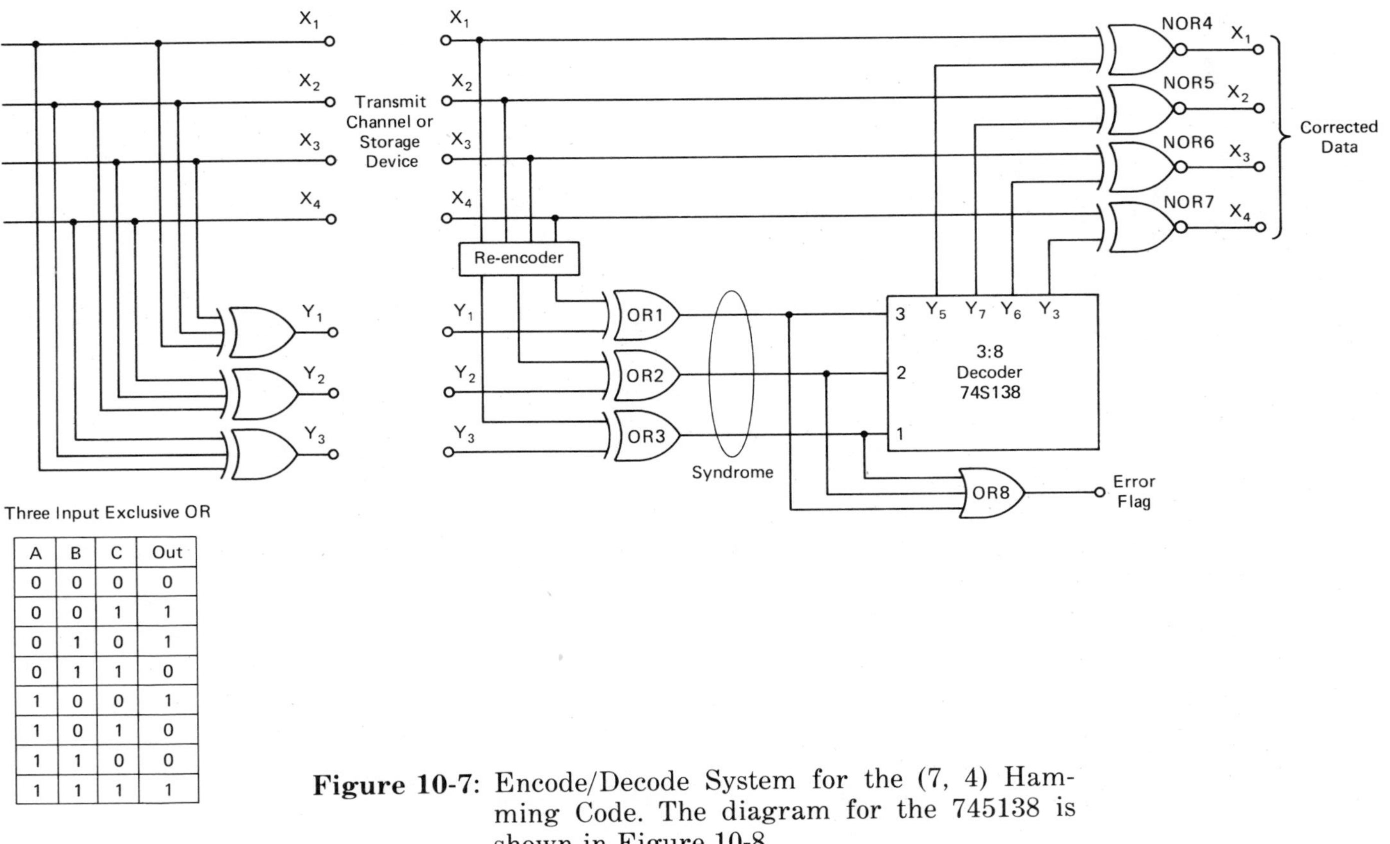

A	B	C	Out
0	0	0	0
0	0	1	1
0	1	0	1
0	1	1	0
1	0	0	1
1	0	1	0
1	1	0	0
1	1	1	1

Figure 10-7: Encode/Decode System for the (7, 4) Hamming Code. The diagram for the 745138 is shown in Figure 10-8.

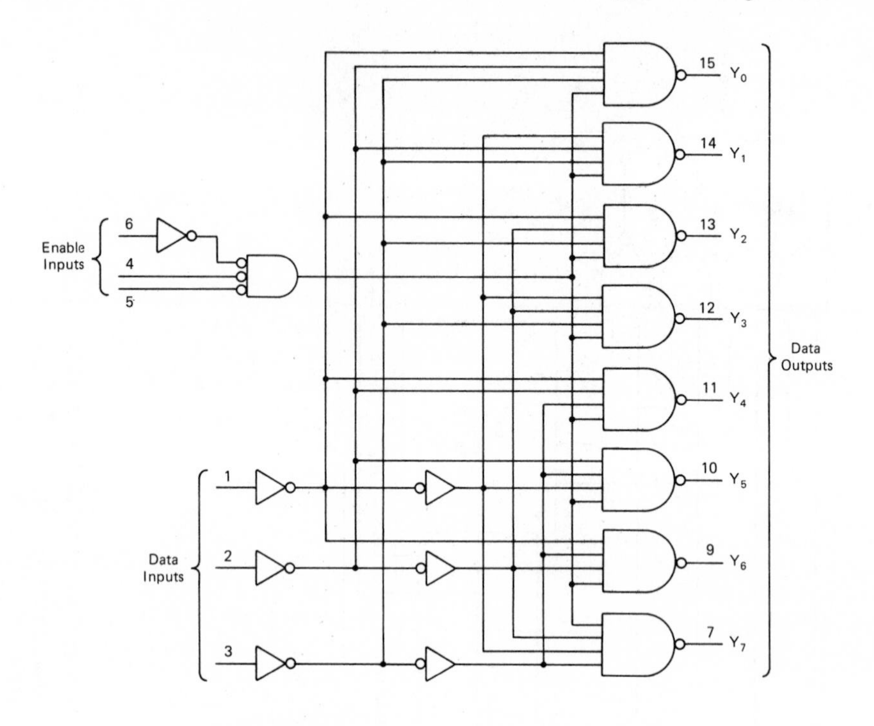

Inputs					Outputs							
Enable		Data										
6	4/5	3	2	1	Y_0	Y_1	Y_2	Y_3	Y_4	Y_5	Y_6	Y_7
X	1	X	X	X	1	1	1	1	1	1	1	1
0	X	X	X	X	1	1	1	1	1	1	1	1
1	0	0	0	0	0	1	1	1	1	1	1	1
1	0	0	0	1	1	0	1	1	1	1	1	1
1	0	0	1	0	1	1	0	1	1	1	1	1
1	0	0	1	1	1	1	1	0	1	1	1	1
1	0	1	0	0	1	1	1	1	0	1	1	1
1	0	1	0	1	1	1	1	1	1	0	1	1
1	0	1	1	0	1	1	1	1	1	1	0	1
1	0	1	1	1	1	1	1	1	1	1	1	0

Figure 10-8: 74S138 3:8 Decoder.

8. The NOR gates will have the following inputs:

NOR 4	0, 1
NOR 5	1, 1
NOR 6	1, 1
NOR 7	1, 0

9. Their outputs will be:

NOR 4	$X_1 = 0$
NOR 5	$X_2 = 1$
NOR 6	$X_3 = 1$
NOR 7	$X_4 = 0$

10. These NOR outputs are the same as the original transmitted data. Therefore, the error has been detected and corrected.

UNDERSTANDING THE G AND H MATRIXES

For the design of encode/decode circuits utilizing Hamming codes, as well as a better understanding of their functions, it is necessary to use matrix algebra. The G and H matrixes completely define the codes in terms of matrix algebra. Either can be used to define a code since when one is known, the other can be derived.

MATRIX ALGEBRA

A matrix is an array of numbers arranged in columns and rows. Figure 10-9 shows a square matrix with four rows

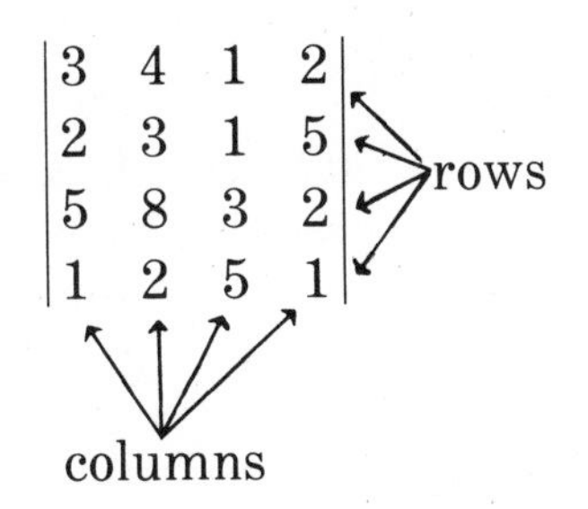

Figure 10-9: Square Matrix.

and four columns. A matrix does not have to be square, it may have any number of rows and/or columns. The manipulation of matrixes is done according to the rules of matrix algebra.

The matrixes used to define Hamming codes may be square or rectangular and consist of ones and zeros. The manipulation of the matrixes is done as follows:

A. The Transpose:

The transpose is defined as interchanging the rows and columns of the matrix.

Example 1:

$$\text{IF } X = \begin{vmatrix} 1 & 0 \\ 1 & 1 \\ 0 & 1 \end{vmatrix}$$

$$X \text{ transpose} = X^t = \begin{vmatrix} 1 & 1 & 0 \\ 0 & 1 & 1 \end{vmatrix}$$

Example 2:

$$\text{IF } \quad Y = \begin{vmatrix} 1 & 0 & 1 & 1 \\ 1 & 1 & 1 & 0 \end{vmatrix}$$

$$Y^t = \begin{vmatrix} 1 & 1 \\ 0 & 1 \\ 1 & 1 \\ 1 & 0 \end{vmatrix}$$

B. Matrix Multiplication

The multiplication of a matrix by a binary number is as follows:

Example 1:

$$0100101 \times \begin{vmatrix} 1 & 0 & 1 \\ 1 & 1 & 1 \\ 1 & 1 & 0 \\ 0 & 1 & 1 \\ 1 & 0 & 0 \\ 0 & 1 & 0 \\ 0 & 0 & 1 \end{vmatrix} = \begin{matrix} 0 \times 1 & 0 \times 0 & 0 \times 1 \\ +1 \times 1 & 1 \times 1 & 1 \times 1 \\ +0 \times 1 & 0 \times 1 & 0 \times 0 \\ +0 \times 0 & 0 \times 0 & 0 \times 0 \\ +1 \times 1 & 1 \times 0 & 1 \times 0 \\ +0 \times 0 & 0 \times 1 & 0 \times 0 \\ +1 \times 0 & 1 \times 0 & 1 \times 1 \\ \hline 0 & 1 & 0 \end{matrix}$$

$$= (\ 0\ 1\ 0\ )$$

The addition in these examples is "mod two" addition. That is:

$$\begin{array}{ll} 0 + 0 = 0 & \\ 0 + 1 = 1 & (1 + 1) = (1 - 1) = 0 \\ 1 + 0 = 1 & \\ 1 + 1 = 0 & \end{array}$$

Example 2:

$$1\ 0\ 1 \times \begin{vmatrix} 1 & 0 & 1 \\ 0 & 1 & 0 \\ 1 & 1 & 1 \end{vmatrix} = \begin{array}{ccc} 1 \times 1 & 1 \times 0 & 1 \times 1 \\ +0 \times 0 & 0 \times 1 & 0 \times 0 \\ +1 \times 1 & 1 \times 1 & 1 \times 1 \\ \hline 0 & 1 & 0 \end{array}$$

G AND H MATRIXES FOR THE (7, 4) CODE

The G matrix is a combination of two matrixes called the I matrix and the P matrix.

$$G = (I,P)$$

The I matrix for the (7, 4) code shown in Figure 10-4 uses messages B, C, E, and I and looks as follows:

$$I_4 = \begin{vmatrix} 1 & 0 & 0 & 0 \\ 0 & 1 & 0 & 0 \\ 0 & 0 & 1 & 0 \\ 0 & 0 & 0 & 1 \end{vmatrix}$$

The P matrix consists of the check bits associated with the data bits of the I matrix.

$$P = \begin{vmatrix} 1 & 0 & 1 \\ 1 & 1 & 1 \\ 1 & 1 & 0 \\ 0 & 1 & 1 \end{vmatrix}$$

Since G =(I,P). The G matrix for the (7, 4) code looks like:

$$G = \left|\begin{array}{cccc} 1 & 0 & 0 & 0 \\ 0 & 1 & 0 & 0 \\ 0 & 0 & 1 & 0 \\ 0 & 0 & 0 & 1 \end{array}\right.\quad\left.\begin{array}{ccc} 1 & 0 & 1 \\ 1 & 1 & 1 \\ 1 & 1 & 0 \\ 0 & 1 & 1 \end{array}\right|$$
$$\underbrace{\qquad}_{I_4}\qquad\underbrace{\qquad}_{P}$$

The G matrix is referred to as the generator matrix since it can be used to generate the entire code. This can be done by the following method.

Example 1:

Generate the (7, 4) code word for the 4-bit message 0110.

$$\boxed{\text{Code word} = \text{message} \times (G)}$$

$$\begin{array}{cccc} \text{Message} & \times & (G) & = \text{Code Word} \\ \downarrow & & \downarrow & \downarrow \end{array}$$

$$0110 \times \left|\begin{array}{c} 1000101 \\ 0100111 \\ 0010110 \\ 0001011 \end{array}\right| = \begin{array}{r} 0 \\ +0100111 \\ +0010110 \\ +\qquad 0 \\ \hline 0110001 \end{array}$$

Desired code word = 0110001

Example 2:

Generate the (7, 4) code word for the 4-bit message 0011.

$$0011 \times \left|\begin{array}{c} 1000101 \\ 0100111 \\ 0010110 \\ 0001011 \end{array}\right| = \begin{array}{r} 0 \\ +\qquad 0 \\ +0010110 \\ +0001011 \\ \hline 0011101 \end{array}$$

Desired code word = 0011101

A quick check of Figure 10-4 will confirm the results of the two examples.

The H matrix is defined as follows:

$$G = (I_4 P)$$

$$H = (P^t I_3)$$

Where: P^t = The transpose of P

For the (7, 4) Hamming code, the H matrix is derived from the G matrix as follows:

$$G = \begin{vmatrix} 1000101 \\ 0100111 \\ 0010110 \\ 0001011 \end{vmatrix} = \overset{I_4}{\begin{vmatrix} 1000 \\ 0100 \\ 0010 \\ 0001 \end{vmatrix}} \quad \overset{P}{\begin{vmatrix} 101 \\ 111 \\ 110 \\ 011 \end{vmatrix}}$$

$$P = \begin{vmatrix} 101 \\ 111 \\ 110 \\ 011 \end{vmatrix} \qquad P^t = \begin{vmatrix} 1110 \\ 0111 \\ 1101 \end{vmatrix}$$

$$H = P^t I_3 = \begin{vmatrix} 1110 & 100 \\ 0111 & 010 \\ 1101 & 001 \end{vmatrix}$$

$$\underbrace{}_{P^t} \quad \underbrace{}_{I_3}$$

The H matrix is important since it gives the method to determine the encode/decode circuits for the Hamming code. The H transpose matrix H^t when multiplied by any code word must equal zero.

$$\boxed{C\,H^t = 0}$$

Example:

Devise an encoder for the (7, 4) Hamming code. (See Figure 10-10.)

Given: A code word $A_1, A_2, A_3, A_4, C_1, C_2, C_3$ and $CH^t = 0$ . .

$$A_1, A_2, A_3, A_4, C_1, C_2, C_3 \times \begin{vmatrix} 101 \\ 111 \\ 110 \\ 011 \\ 100 \\ 010 \\ 001 \end{vmatrix} = 0$$

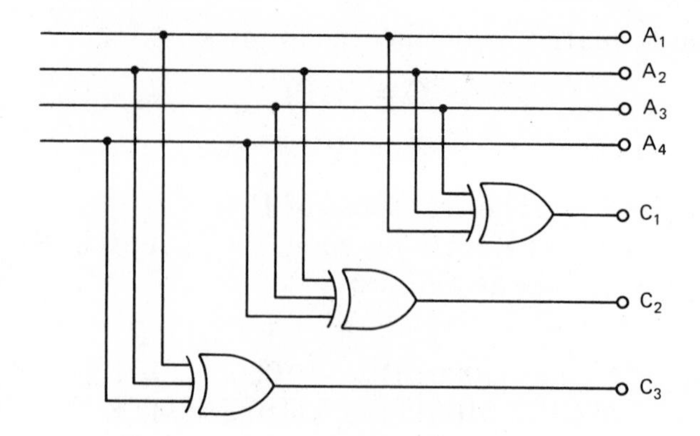

Figure 10-10: An Encoder for a (7, 4) Hamming Code.

$$(A_1 \times 1) + (A_2 \times 1) + (A_3 \times 1) + (C_1 \times 1) = 0$$

$$\boxed{C_1 = A_1 + A_2 + A_3}$$

$$(A_2 \times 1) + (A_3 \times 1) + (A_4 \times 1) + (C_2 \times 1) = 0$$

$$\boxed{C_2 = A_2 + A_3 + A_4}$$

$$(A_1 \times 1) + (A_2 \times 1) + A_4 \times 1) + (C_3 \times 1) = 0$$

$$\boxed{C_3 = A_1 + A_2 + A_4}$$

USING HAMMING CODES ON LARGER DATA BLOCKS

Hamming codes exist over a wide range of data lengths. In general:

$$\boxed{N = 2^{(N - K)} - 1}$$

Where: N = total number of bits in the code word.
$(N - K)$ = number of check bits.

N − K	N	
1	1	
2	3	⟶(3, 1)
3	7	⟶(7, 4)—4-bit data blocks
4	15	⟶(15, 11)—8-bit data blocks
5	31	⟶(31, 26)—16-bit data blocks
6	63	⟶(63, 57)—32-bit data blocks

The 7, 4 code is very convenient because it applies directly to 4-bit data blocks. There are no Hamming codes that correspond directly to 8-, 16-, or 32-bit data blocks, so it is common to use the code for the next larger data block. When this is done it is necessary to "shorten" the code to comply to the number of bits being used in the block.

How to Shorten a Code

Figure 10-11 gives the G and H matrixes for the (15, 11) Hamming code. This is the commonly used code for 8-bit data blocks. In order to use this code for 8-bit data, it is necessary to shorten the code, that is, you need to change the (15, 11) code to a (12, 8) code. When the code is shortened, the data as well as the total message is reduced by three bits. The total number of check bits remains at four.

To shorten the code, the G and H matrixes are changed as shown in Figure 10-12. The H matrix has the first three columns removed. The G matrix is reduced as shown in the figure. To encode the shortened code, the encoding method for the (15, 11) code is used except that the inputs from the "missing" data bits are not used. Figure 10-13 shows the encoding scheme. The + symbol indicates modulus two addition or a multiple input exclusive OR.

$$G = \begin{vmatrix} 1000101 \\ 0100111 \\ 0010110 \\ 0001011 \end{vmatrix} \qquad H = \begin{vmatrix} 1110100 \\ 0111010 \\ 1101001 \end{vmatrix}$$

7, 4 HAMMING

$$G = \begin{vmatrix} 100000000001001 \\ 010000000001101 \\ 001000000001111 \\ 000100000001110 \\ 000010000000111 \\ 000001000001010 \\ 000000100000101 \\ 000000010001011 \\ 000000001001100 \\ 000000000100110 \\ 000000000010011 \end{vmatrix} = I_{11}P$$

$$H = \begin{vmatrix} 111101011001000 \\ 011110101100100 \\ 001111010110010 \\ 111010110010001 \end{vmatrix} = P^{t}I_{4}$$

15, 11 HAMMING

Figure 10-11: G and H Matrixes for Common (7, 4) and (15, 11) Hamming Codes.

$$G(15, 11) = \left| \begin{array}{c} 100000000001001 \\ 010000000001101 \\ 001000000001111 \\ \hline 000100000001110 \\ 000010000000111 \\ 000001000001010 \\ 000000100000101 \\ 000000010001011 \\ 000000001001100 \\ 000000000100110 \\ 000000000010011 \end{array} \right| \qquad G(12, 8) = \left| \begin{array}{c} 100000001110 \\ 010000000111 \\ 001000001010 \\ 000100000101 \\ 000010001011 \\ 000001001100 \\ 000000100110 \\ 000000010011 \end{array} \right|$$

$$H(15, 11) = \left| \begin{array}{c} 111101011001000 \\ 011110101100100 \\ 001111010110010 \\ 111010110010001 \end{array} \right| \qquad H(12, 8) = \left| \begin{array}{c} 101011001000 \\ 110101100100 \\ 111010110010 \\ 010110010001 \end{array} \right|$$

Figure 10-12

$$CH^t = 0 \qquad H^t = \left| \begin{array}{c} 1110 \\ 0111 \\ 1010 \\ 0101 \\ 1011 \\ 1100 \\ 0110 \\ 0011 \\ 1000 \\ 0100 \\ 0010 \\ 0001 \end{array} \right|$$

$$C_1 = A_1 + A_3 + A_5 + A_6$$

$$C_2 = A_1 + A_2 + A_4 + A_6 + A_7$$

$$C_3 = A_1 + A_2 + A_3 + A_5 + A_7 + A_8$$

$$C_4 = A_2 + A_4 + A_5 + A_8$$

Figure 10-13

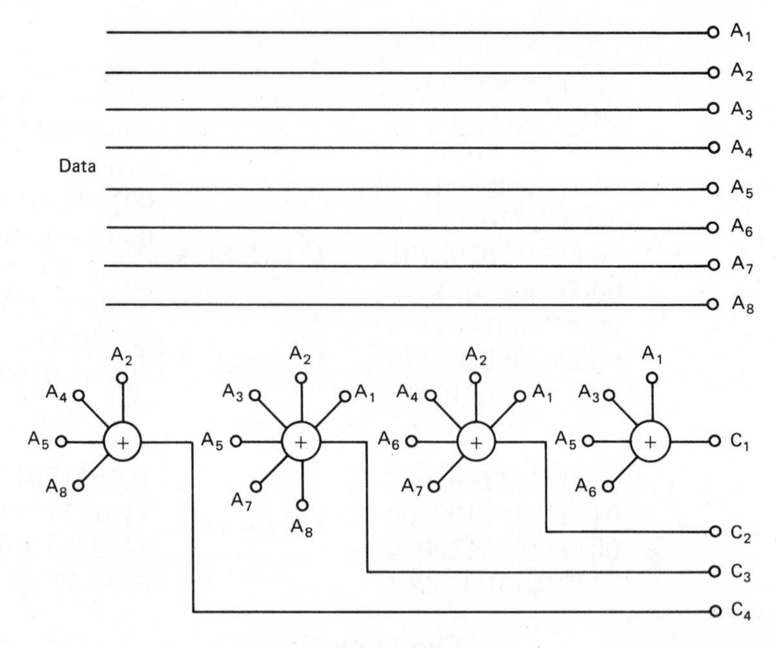

Figure 10-13 (continued)

How to Extend Hamming Codes

It is possible to add an extra check bit to a Hamming code that will increase the minimum distance. This can increase the correction/detection capability of the code.

Consider the G matrix for the (7, 4) Hamming code Figure 10-14. If we add an extra check bit as shown, it can produce 8-bit messages with a minimum distance of four. The extra check bit also produces even parity in all messages.

How to Avoid Pattern Sensitivity

The 7, 4 Hamming code in Figure 10-4 contains the all zero message (A) and the all ones message (P). These patterns can be caused by several conditions, loss of power, removal of a PC card, and others. Since these are valid code words, the decoding will not detect these conditions as errors. This condition is referred to as pattern sensitivity.

$$G = \begin{vmatrix} 1000101 \\ 0100111 \\ 0010110 \\ 0001011 \end{vmatrix} \qquad G \text{ extended} = \begin{vmatrix} 10001011 \\ 01001110 \\ 00101101 \\ 00010111 \end{vmatrix}$$

(7, 4) Hamming	Extended Hamming, (8, 4)	
0000 000	0000 0000	
0001 011	0001 0111	
0010 110	0010 1101	
0011 101	0011 1010	
0100 111	0100 1110	
0101 100	0101 1001	
0110 001	0110 0011	
0111 010	0111 0100	Even
1000 101	1000 1011	Parity
1001 110	1001 1100	
1010 011	1010 0110	
1011 000	1011 0001	
1100 010	1100 0101	
1101 001	1101 0010	
1110 100	1110 1000	
1111 111	1111 1111	

(7, 4) Hamming
minimum distance = 3

Extended Hamming, (8, 4)
Minimum distance = 4

Figure 10-14

To avoid this condition, the circuit shown in Figure 10-15 is used. The check bits are inverted prior to being stored and inverted back again before the message is decoded. With this circuit, the message words at the storage device never contain the all zero or all ones words. The conditions listed above that can cause these conditions will now be detected as errors.

HAMMING CODES IN PRACTICE

With the use of 16K and 64K RAM memory chips, the error rate has risen by several orders of magnitude (see Table

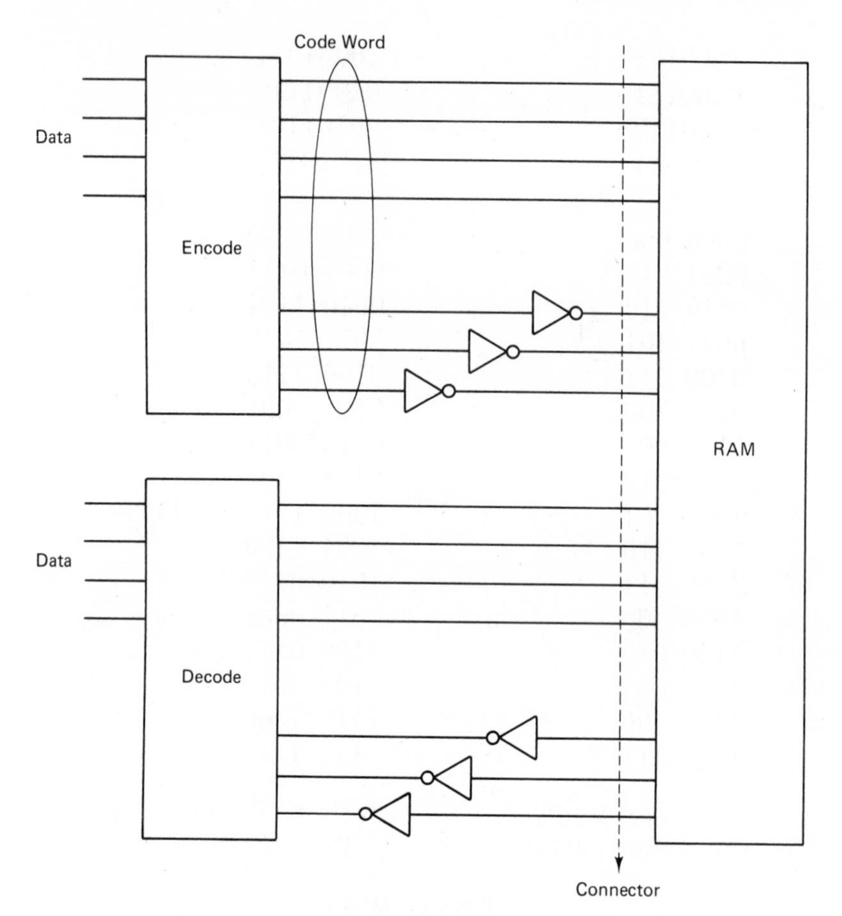

Figure 10-15

10-1). With the smaller density 1K devices, the probability of error was so small that it could be ignored. With the 64K devices the probability of error has grown to the point where something must be done to detect and correct these errors.

The errors that occur in solid-state memory devices can be placed in two main categories: (1) hard errors and (2) soft errors. The hard errors are errors that occur every time a certain address is read. They are usually a defect in the device itself and can be handled by replacement of the device or by mapping techniques so that certain memory addresses will not be used. A soft error is one that occurs at random. It can be caused by power interruptions, noise spikes, or by the "alpha" radiation problem. The soft errors

Density Bits/Chip	Typical Error Rate (% per 1,000 Hours)	
	Soft*	Hard**
1K	.001	.0001
4K	.02	.002
16K	.10	.011
64K	.5***	.016

*Reflects alpha particles only. Does not include errors due to noise, power, patterns.
**After infant mortality.
***Based on initial customer evaluation.
Note: 0.1% per 1000 hours equals 1 failure in 10^6 hours.

Table 1. Errors are Increasing,

are the ones which need to be handled by error correction codes.

A scheme for error correction in RAM circuits is shown in Figure 10-16. This circuit is a generalization of the encoding/decoding scheme shown in Figure 10-7. It is common to use extended Hamming codes with this circuit so that the minimum distance is four.

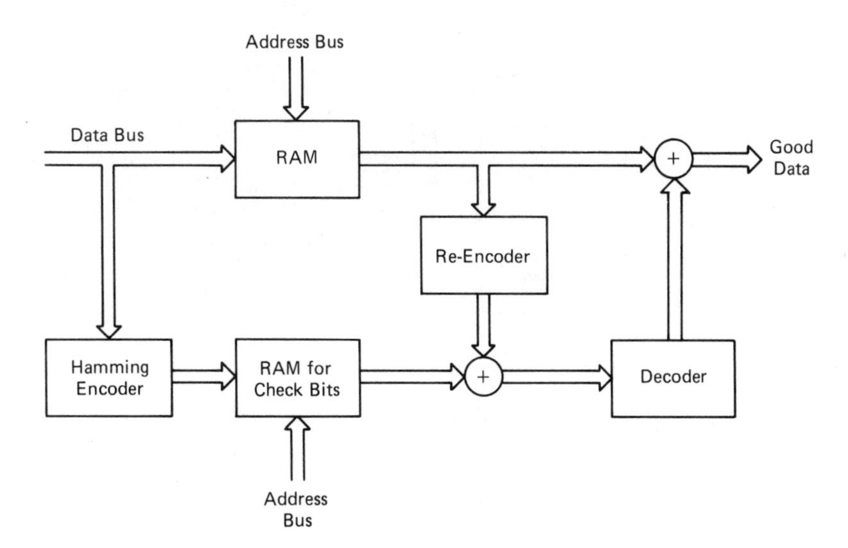

Figure 10-16

Along with standard error correction while using 16K and 64K chips, it has become common to have a memory refresh cycle. The circuits contain control logic which (at periodic intervals) dump the data from RAM, cycle the data through error detection and correction circuits, and re-store the data in the RAM. Several LSI chips have been designed and manufactured to handle the entire support (error correction and refresh cycle) of large storage RAMs. One such system is manufactured by "advanced micro devices." The block diagram of the complete system is shown in Figure 10-17. The error detection and correction chip itself is the AMD 2960. This device is capable of encoding and decoding shortened Hamming codes. In can handle data lengths of 8, 16, 32, and 64 bits by proper connection of the chip. Pin connections and a block diagram of the chips' function are shown in Figures 10-18 to 10-19.

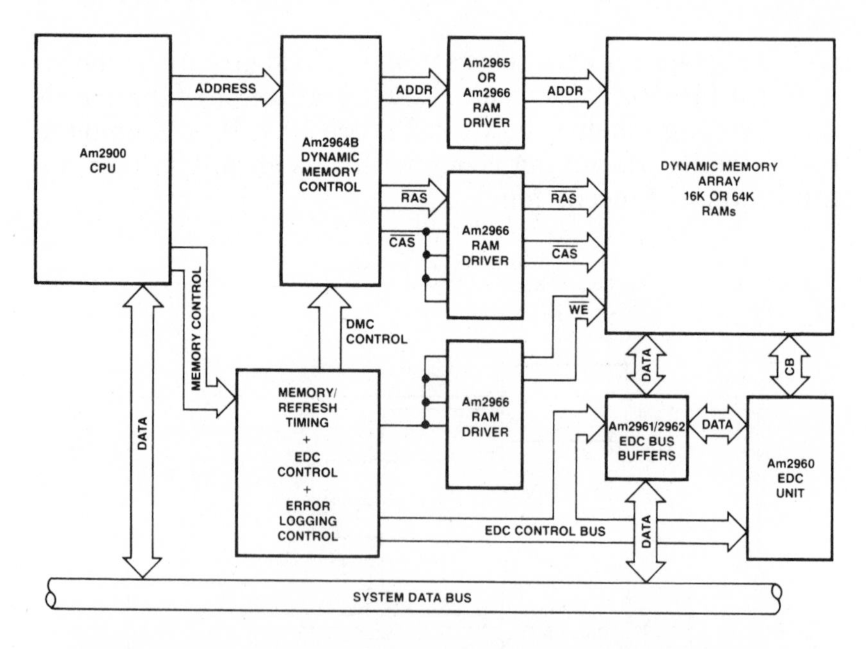

Figure 10-17: Am2900 High Performance Computer Memory. Copyright © 1984 Advanced Micro Devices, Inc. Reproduced with permission of copyright owner. All rights reserved.

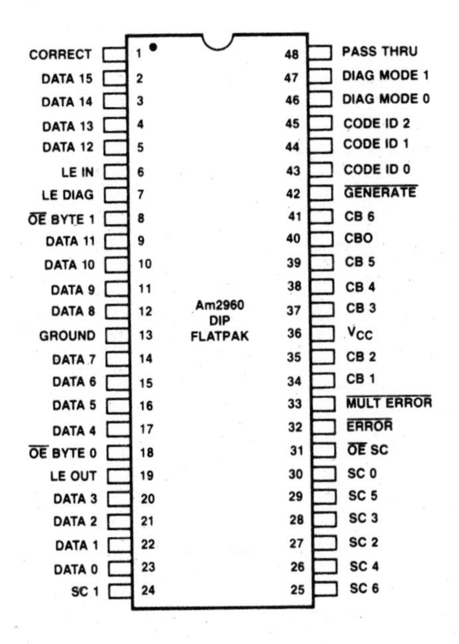

Note: Pin 1 is marked for orientation.

Figure 10-18: Connection Diagram (Top View).

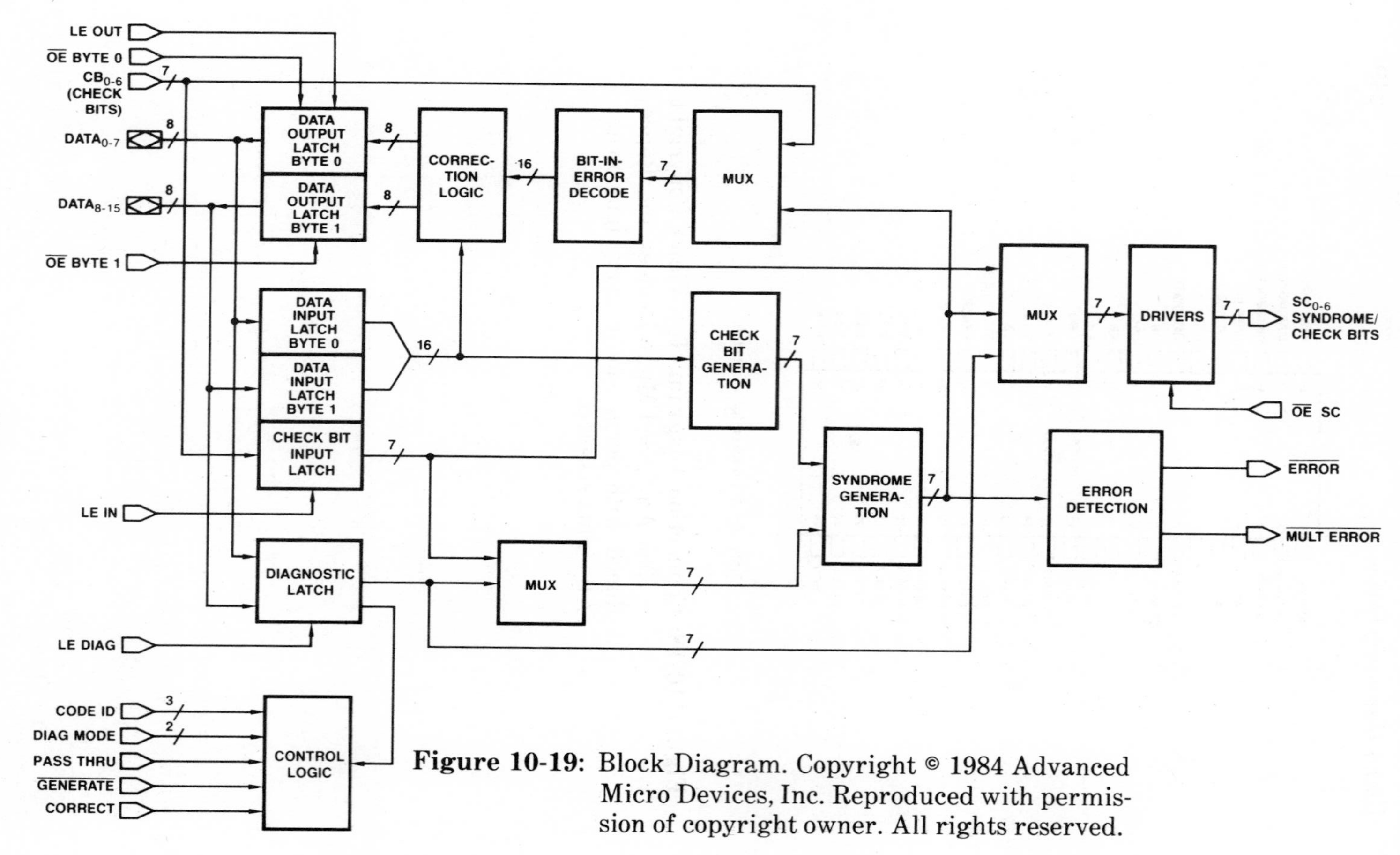

Figure 10-19: Block Diagram.

11

Understanding Binary Cyclic Codes

INTRODUCTION

Parity checks and Hamming codes discussed in chapters 9 and 10 are examples of block codes. They work well with RAMS and other devices that handle 4-, 8-, and 16-bit binary data busses.

When memory devices such as magnetic tape or magnetic disks are used it is common to see binary cyclic codes used. These devices store large strings of binary data and their input/output is usually in serial form rather than parallel. It is possible to use parity and/or Hamming codes on tape and disks devices. Indeed, the (7, 4) Hamming code can be expressed in cyclic form. But, it is more common to see these codes defined in terms of a binary polynomial called a generator polynomial.

To understand the action of a binary cyclic code, it is necessary to understand binary polynomials and how to work with them.

DEFINITION OF MOD 2 ARITHMETIC

Addition and subtraction in mod 2 is defined as follows:

$$0 + 0 = 0$$

$$1 + 0 = 1$$

$$0 + 1 = 1$$

$$1 + 1 = 0$$

This is the function of a binary half-adder or an exclusive OR gate. Also, addition and subtraction in mod 2 are identical operations.

$$+1 = -1$$
$$+X = -X$$
$$1 + 1 = 0$$
$$1 - 1 = 0$$

The output of a half-adder is the same as the output of a half-subtractor.

WORKING WITH BINARY POLYNOMIALS

A polynomial is the summation of various powers of variables. A coefficient is attached to each variable. In general, a polynomial can be written:

$$G(X) = AX^2 + BX^1 + CX^0$$

In the case of binary polynomials, the coefficients (A,B,C) can have only two values. They must be either zero or one. The general binary polynomial can be written:

$$G(X) = X^2 + X^1 + 1$$

Binary polynomials can represent binary numbers directly as follows:

BINARY NUMBER		BINARY POLYNOMIAL
10110	$\longrightarrow$	$1X^4 + 0X^3 + 1X^2 + 1X^1 + 0X^0$
		$= X^4 + X^2 + X$
11011	$\longrightarrow$	$1X^4 + 1X^3 + 0X^2 + 1X^1 + 1X^0$
		$= X^4 + X^3 + X + 1$
01010	$\longrightarrow$	$0X^4 + 1X^3 + 1X^2 + 1X^1 + 0X^0$
		$= X^3 + X$
11111	$\longrightarrow$	$1X^4 + 1X^3 + 1X^2 + 1X^1 + 1X^0$
		$= X^4 + X^3 + X^2 + X + 1$

Binary polynomials can be manipulated the same as binary numbers using the rules of mod 2 arithmetic.

Example: Add the following polynomials

$$\begin{array}{r} X^4 + X^3 + X + 1 \\ + X^4 + X^2 + X + 0 \\ \hline 0 + X^3 + X^2 + 0 + 1 \end{array}$$

$$= (X^3 + X^2 + 1)$$

since $(X^4 + X^4) = (X^4 - X^4) = 0$

Example:

Divide $(X^8 + X^5 + X^3 + 1)$ by $(X^3 + X + 1)$

$$\begin{array}{r l} & X^5 + X^3 + X \\ \hline X^3 + X + 1 \,) & X^8 + X^5 + X^3 + 1 \\ & X^8 + X^6 + X^5 \\ \hline & X^6 + X^3 + 1 \\ & X^6 + X^4 + X^3 \\ \hline & X^4 + 1 \\ & X^4 + X^2 + X \\ \hline & X^2 + X + 1 \end{array}$$

Example:

Divide $(X^4 + X^2)$ by $(X^2 + 1)$

$$\begin{array}{r l} & X^2 \\ \hline X^2 + 1 \,) & X^4 + X^2 \\ & X^4 + X^2 \\ \hline & 0 \ + 0 \end{array}$$

Using the rules of mod 2 arithmetic, any arithmetic operation can be done on binary polynomials.

DEFINITION OF CYCLIC CODES

Assume that a large block of data bits is being transmitted or being stored on magnetic tape. The binary cyclic code (also called cyclic redundancy check) is generated by passing the data bits through a special circuit during transmission or storage and then adding them on at the end of the message. This is shown in Figure 11-1.

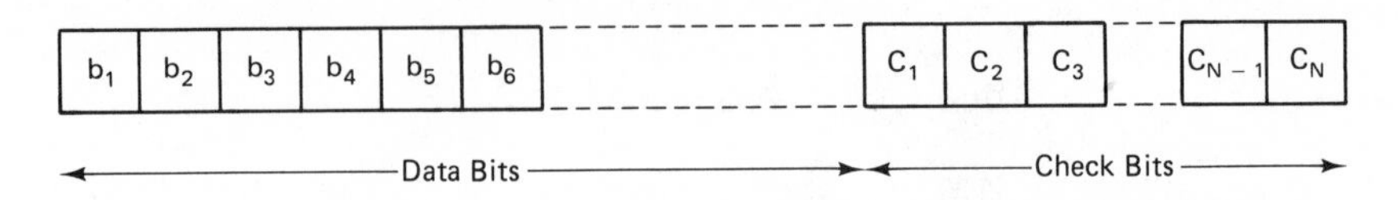

Figure 11-1

The binary data bits can be represented by a binary polynomial (P). If this polynomial (P) is divided by a "predetermined" generator polynomial (G), the result will be a quotient plus a remainder.

$$\frac{P}{G} = \text{quotient} + \text{remainder}$$

The check bits added to the end of the data, are the remainder from the division of the (P) and (G) polynomials. The circuit that generates the check bits performs a continuous division of the data bit polynomial by the generator polynomial.

Example:

The transmitted data is a six-bit binary number. Check bits are to be added according to the generating polynomial $G(X) = 1 + X + X^3$

Find the check bits.

A. Transmit data = 101101 = $1 + X^2 + X^3 + X^5$

B. The total transmitted message will have nine total bits since there will be three check bits added. Each data bit will then be increased by X^3 in its position.

$$(X^3) \cdot (1 + X^2 + X^3 + X^5) = X^3 + X^5 + X^6 + X^8$$

C. The data polynomial is divided by the generator polynomial. The remainder will be added as check bits.

$$
\begin{array}{l}
\qquad\qquad\qquad X^5 + 1 \longrightarrow \text{quotient} \\
\underbrace{X^3 + X + 1}_{\text{Generator Polynomial}} \;\big)\; X^8 + X^6 + X^5 + X^3 \longrightarrow \text{(data polynomial)} \\
\qquad\qquad\qquad \underline{X^8 + X^6 + X^5} \\
\qquad\qquad\qquad\qquad\qquad X^3 \\
\qquad\qquad\qquad\qquad\qquad \underline{X^3 + X + 1} \\
\qquad\qquad\qquad\qquad\qquad\quad X + 1 = \text{(remainder)}
\end{array}
$$

D. The entire transmitted message (data bits + check bits) will be:

$$
\begin{array}{c|c}
X^8 + X^6 + X^5 + X^3 & + X + 1 \\
\underbrace{1\;\;0\;\;1\;\;1\;\;0\;\;1}_{\text{data bits}} & \underbrace{0\;\;1\;\;1}_{\text{check bits}}
\end{array}
$$

E. When the message is received or read back, the entire polynomial is again divided by the generator polynomial. If there are no errors, the remainder will be zero. If an error occurs in either the data bits or check bits, the remainder will be non-zero.

$$
\begin{array}{l}
\qquad\qquad\qquad X^5 + 1 \\
\text{Generator Polynomial } X^3 + X + 1 \;\big)\; X^8 + X^6 + X^5 + X^3 + X + 1 \longrightarrow \left\{\text{message received correctly}\right. \\
\qquad\qquad\qquad \underline{X^8 + X^6 + X^5} \\
\qquad\qquad\qquad 0 \quad 0 \quad 0 + X^3 + X + 1 \\
\qquad\qquad\qquad\qquad\qquad \underline{X^3 + X + 1} \\
\qquad\qquad\qquad\qquad\qquad\quad 0 \searrow \left\{\text{zero remainder}\right.
\end{array}
$$

F. When the message is received with an error, the remainder will be non-zero. Assume that the error caused the following message to be received; an error in the third data bit from the left.

$$
\underbrace{1\;\;0\;\;0\;\;1\;\;0\;\;1}_{\text{data bits}} \qquad \underbrace{0\;\;1\;\;1}_{\text{check bits}}
$$

$(X^8 + X^5 + X^3 + X + 1)$

$$
\begin{array}{r}
X^5 + 1 \\
X^3 + X + 1 \;\big)\; X^8 + X^5 + X^3 + X + 1 \quad \text{received "wrong" message} \\
X^8 + X^6 + X^5 \\
\hline
X^6 + X^3 + X + 1 \\
X^3 + X + 1 \\
\hline
X^6 \quad \text{remainder}
\end{array}
$$

Binary cyclic codes are generated by dividing the message bits by a "pre-determined" binary polynomial. At the end of the message bits, the bits formed by the remainder of this division are added to the message as check bits. When the message is received/read back, the message is again divided by the same binary polynomial. Since the remainder has been added, the result of the second division should have a remainder of zero. If the remainder is not zero, an error has occurred. This error can exist in the data bits, the check bits, or both.

EXAMPLE OF CYCLIC CODE ACTION

The mathematics of binary polynomials is accomplished in physical circuits by the use of shift registers and exclusive OR gates. The exclusive OR function is the same as binary addition and is usually shown in circuits as:

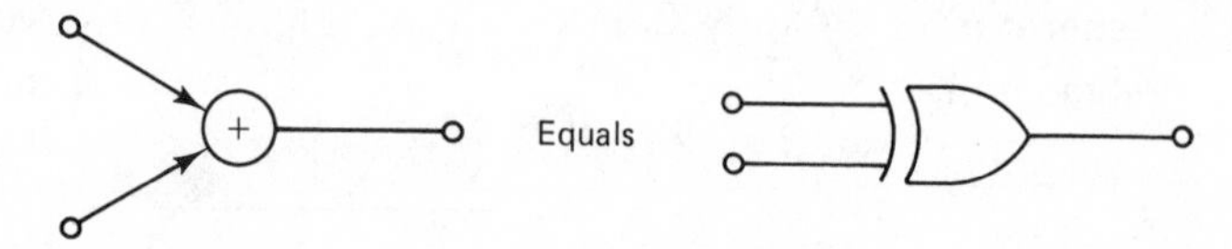

A circuit that can generate cyclic codes for the generator polynomial $G(X) = 1 + X + X^3$ is shown in Figure 11-2. Assuming four-bit data messages, a total of seven bits will be transmitted as follows:

A. Switch S_1 is in position A so that the data bits are transmitted while at the same time they are being shifted through the register.

B. At the end of the data bits, switch S_1 is placed in

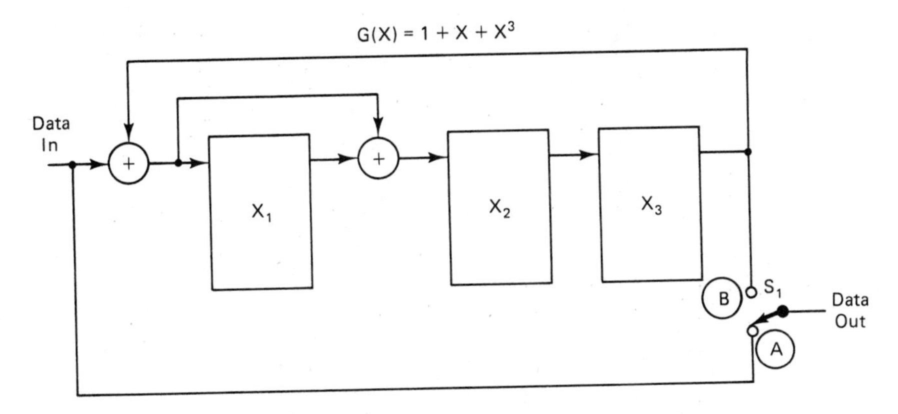

Let: Data = 1 1 0 1

Data	Shift	X_1	X_2	X_3
		0	0	0
1	1	1	1	0
1	2	1	0	1
0	3	1	0	0
1	4	1	0	0

Figure 11-2: Circuit to Generate a Binary Cyclic Code for the Polynomial $G(x) = 1 + x + x^3$ (Total Message 1 1 0 1 0 0 1).

position B and the contents of the shift register are transmitted.

C. Figure 11-2 shows the action of the shift register when the data message 1101 is shifted through MSB first. At the end of four shifts, the register contains $X_1 = 1$, $X_2 = 0$, $X_3 = 0$.

D. The total transmitted message is then 1101001.

E. A complete list of the messages generated by this circuit is shown in Figure 11-3. Note that this is the same as the (7, 4) Hamming code discussed in Chapter 10.

F. When the message is received/read back, the same circuit is used to check for errors. Figure 11-4 shows the message 1101001 being shifted back through the register that generated the code. At the end of seven shifts, the register contains zero indicating no errors. Had any errors been present, the register would have been non-zero.

Data	Codes	Total Message
0 0 0 0	0 0 0	0 0 0 0 0 0 0
0 0 0 1	0 1 1	0 0 0 1 0 1 1
0 0 1 0	1 1 0	0 0 1 0 1 1 0
0 0 1 1	1 0 1	0 0 1 1 1 0 1
0 1 0 0	1 1 1	0 1 0 0 1 1 1
0 1 0 1	1 0 0	0 1 0 1 1 0 0
0 1 1 0	0 0 1	0 1 1 0 0 0 1
0 1 1 1	0 1 0	0 1 1 1 0 1 0
1 0 0 0	1 0 1	1 0 0 0 1 0 1
1 0 0 1	1 1 0	1 0 0 1 1 1 0
1 0 1 0	0 1 1	1 0 1 0 0 1 1
1 0 1 1	0 0 0	1 0 1 1 0 0 0
1 1 0 0	0 1 0	1 1 0 0 0 1 0
1 1 0 1	0 0 1	1 1 0 1 0 0 1
1 1 1 0	1 0 0	1 1 1 0 1 0 0
1 1 1 1	1 1 1	1 1 1 1 1 1 1

Figure 11-3

UNDERSTANDING PRIMITIVE POLYNOMIALS

Binary cyclic codes are almost always generated by what are known as primitive polynomials. The definition of primitive polynomials is beyond the mathematical scope of this book but the following properties pertain.

1. For any number M there is at least one primitive polynomial.
2. Primitive polynomials reduce the possibility of an error going undetected.

With an arbitrary binary polynomial being used, it is possible to devise error patterns that cannot be detected because the polynomial will divide the error evenly without any remainder.

A list of primitive polynomials up through degree 20 is shown in Figure 11-5.

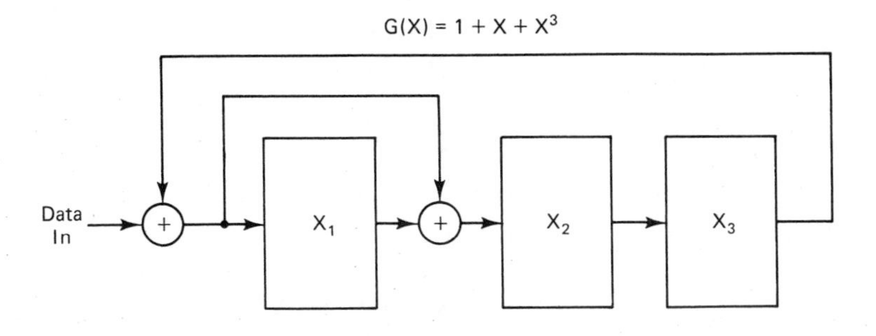

Data	Shift #	X_1	X_2	X_3
		0	0	0
1	1	1	1	0
1	2	1	0	1
0	3	1	0	0
1	4	1	0	0
0	5	0	1	0
0	6	0	0	0
1	7	0	0	0

Figure 11-4: When the message 1101001 is read back through the shift register, the register contains zero, indicating that no error occurred.

EXAMPLES OF CIRCUIT ACTION WHEN AN ERROR OCCURS

To understand the detection of an error, it is necessary to trace through the shift register when the message contains an error. Consider the following conditions:

A. The data bits 1000 were transmitted. The check bits were generated by the polynomial $1 + X + X^3$, the same as Figures 11-2 and 11-6.

B. From Figure 11-3 the check bits generated were 101 so that the total message was 1000101.

C. When the message was received, there was an error. The message received was 1000001.

Degree	Primitive Polynomial
3	$1 + X + X^3$
4	$1 + X + X^4$
5	$1 + X^2 + X^5$
6	$1 + X + X^6$
7	$1 + X^3 + X^7$
8	$1 + X^2 + X^3 + X^4 + X^8$
9	$1 + X^4 + X^9$
10	$1 + X^3 + X^{10}$
11	$1 + X^2 + X^{11}$
12	$1 + X + X^4 + X^6 + X^{12}$
13	$1 + X + X^3 + X^4 + X^{13}$
14	$1 + X + X^6 + X^{10} + X^{14}$
15	$1 + X + X^{15}$
16	$1 + X + X^3 + X^{12} + X^{16}$
17	$1 + X^3 + X^{17}$
18	$1 + X^7 + X^{18}$
19	$1 + X + X^2 + X^5 + X^{19}$
20	$1 + X^3 + X^{20}$

Figure 11-5: Primitive Polynomials.

D. The results of this message being shifted through the register are shown in Figure 11-6. After the final shift, the register is non-zero which indicates an error.

E. The data left in the register is known as the syndrome. An all-zero syndrome always indicates error-free.

F. The syndrome can be used to correct the error. Note that the syndrome contains three binary digits so that there are eight possible combinations of these digits (000-111). The all-zero syndrome indicates error-free; this leaves seven combinations. Each of these combinations corresponds to one of the seven binary digits in the message.

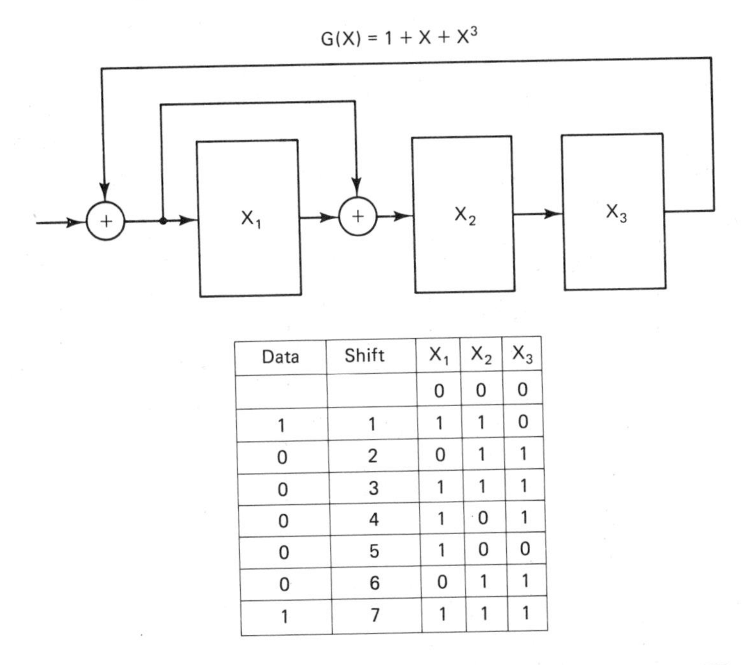

Data	Shift	X_1	X_2	X_3
		0	0	0
1	1	1	1	0
0	2	0	1	1
0	3	1	1	1
0	4	1	0	1
0	5	1	0	0
0	6	0	1	1
1	7	1	1	1

Figure 11-6: The transmitted message was 1000101. The received message was 1000001. Therefore an error was detected, and the register does not contain zero.

EXAMPLES OF ENCODING AND DECODING

To divide by binary polynomials, shift registers can take several forms. It is possible to see a pattern in the construction of these registers which are known as LFSR (Linear Feedback Shift Register).

Shift registers for various polynomials are shown in Figure 11-7. These registers are known as internal XOR registers since the exclusive ORs are internal to the register. It is possible to construct external XOR shift registers also. Figure 11-8 shows both internal and external XOR shift registers to accomplish division by the polynomial $X^6 + X^5 + X^4 + X^3 + 1$.

Cyclic codes have been used on large magnetic data storage systems for many years. The generator polynomials

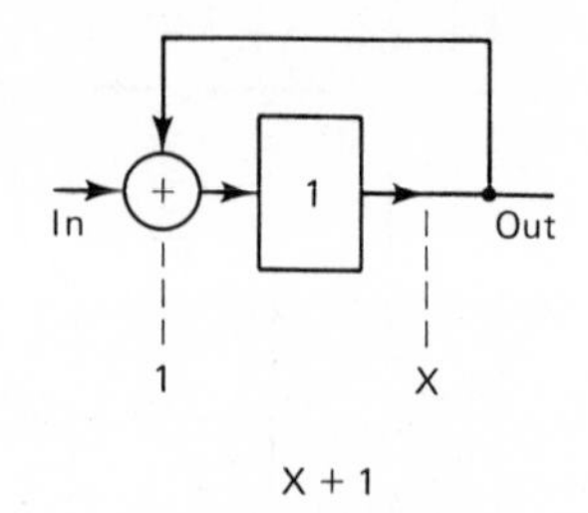
In
1
Out
X
X + 1

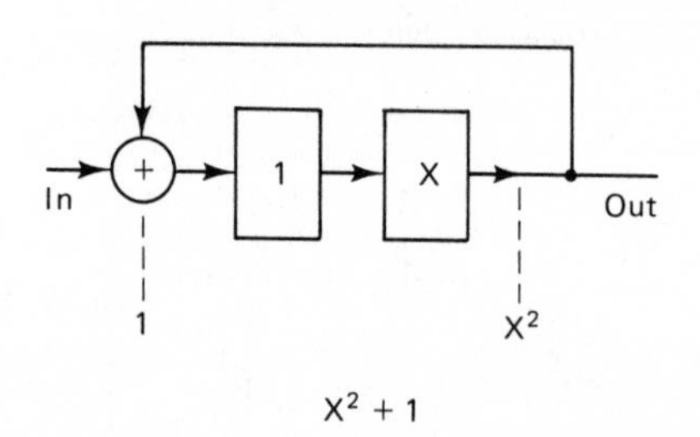
In
1
X
Out
1
X²
X² + 1

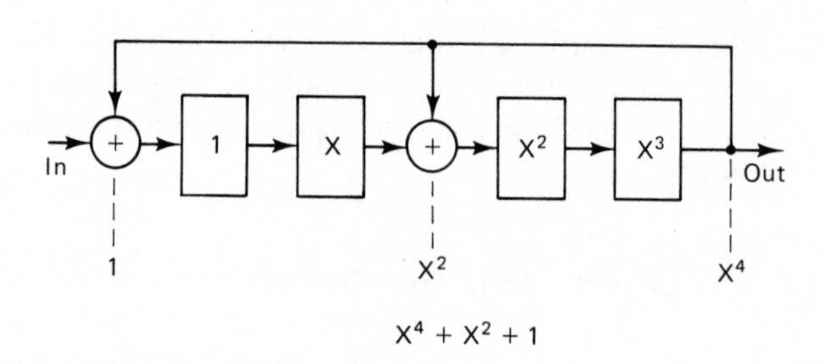
In
1
X
X²
X³
Out
1
X²
X⁴
X⁴ + X² + 1

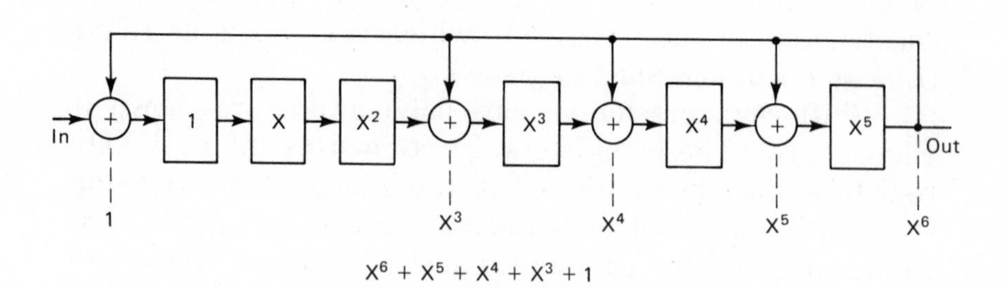
In
1
X
X²
X³
X⁴
X⁵
Out
1
X³
X⁴
X⁵
X⁶
X⁶ + X⁵ + X⁴ + X³ + 1

Figure 11-7

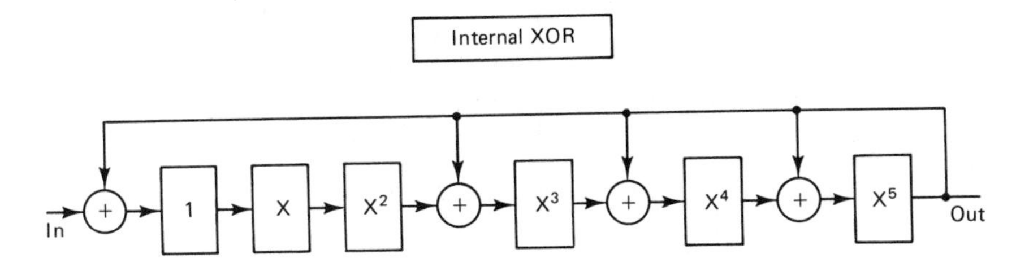

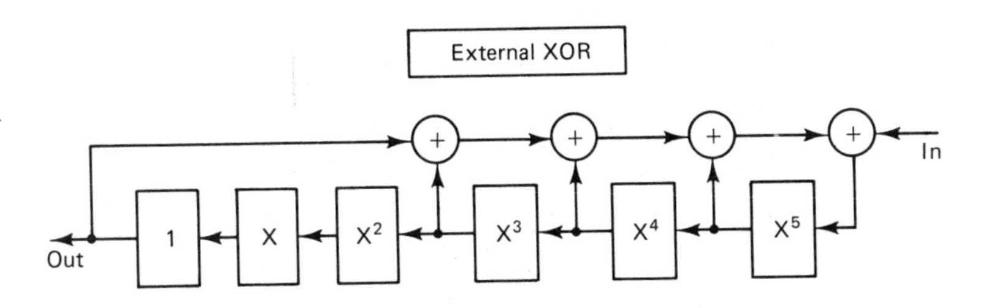

Figure 11-8: Two different ways to divide by $x^6 + x^5 + x^4 + x^3 + 1$. XOR = exclusive OR gate.

are quite large since they are used on large data fields and they must be able to correct errors that are several bits long.

Three of these codes are shown in Figure 11-9. Figure 11-10 shows the linear feedback shift register (LSFR) used to implement the 56 bit code shown in Figure 11-9. The 56-bit code can correct errors up to and including 11 bits. If there is an error on the disc greater than 11 bits long, the code will detect the error, but will not be able to correct it. In the event that a disc has an uncorrectable error, it is handled by other means. An alternate track can be assigned or the error can be skipped by writing sync bytes before and after the defect.

The polynomials given in Figure 11-9 are so common that the electronics needed to encode/decode have been put on a single IC chip. The ADVANCED MICRODEVICES Am Z8065 is such a chip. It can encode/decode any one of the three polynomials simply by proper input to the S_0, S_1 inputs. A pin connection and functional block diagram of this chip is shown in Figure 11-11.

Polynomial	Length in Bits	Data Length in Bits	Burst Correction Capability in Bits	Comments
$X^{56} + X^{55} + X^{49} + X^{45} + X^{41} + X^{39} + X^{38} + X^{37} + X^{36} + X^{31} + X^{22} + X^{19} + X^{17} + X^{16} + X^{15} + X^{14} + X^{12} + X^{11} + X^{9} + X^{5} + 1$	56	585,442	11	Used on I.B.M. 3330 Disc Drives
$X^{32} + X^{23} + X^{21} + X^{11} + X^{2} + 1$	32	42,987	11	Used on D.E.C. Disc Drives
$X^{48} + X^{36} + X^{35} + X^{23} + X^{21} + X^{15} + X^{13} + X^{8} + X^{2} + 1$	48	—	7	Used on S.T.C. 8350 Disc Drive

Figure 11-9: Examples of Commonly Used Generator Polynomials on Large Magnetic Disc Data Storage Systems.

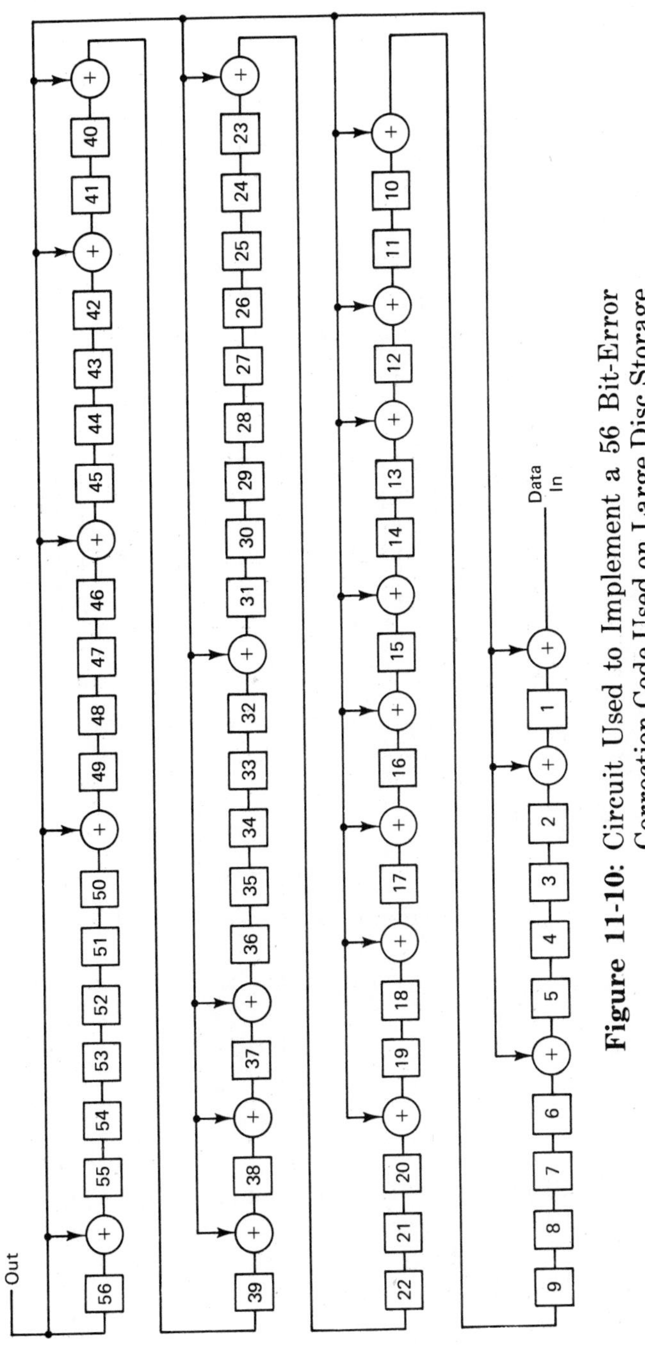

Figure 11-10: Circuit Used to Implement a 56 Bit-Error Correction Code Used on Large Disc Storage Devices. This Code Can Correct Errors Up to 11 Bits in Length.

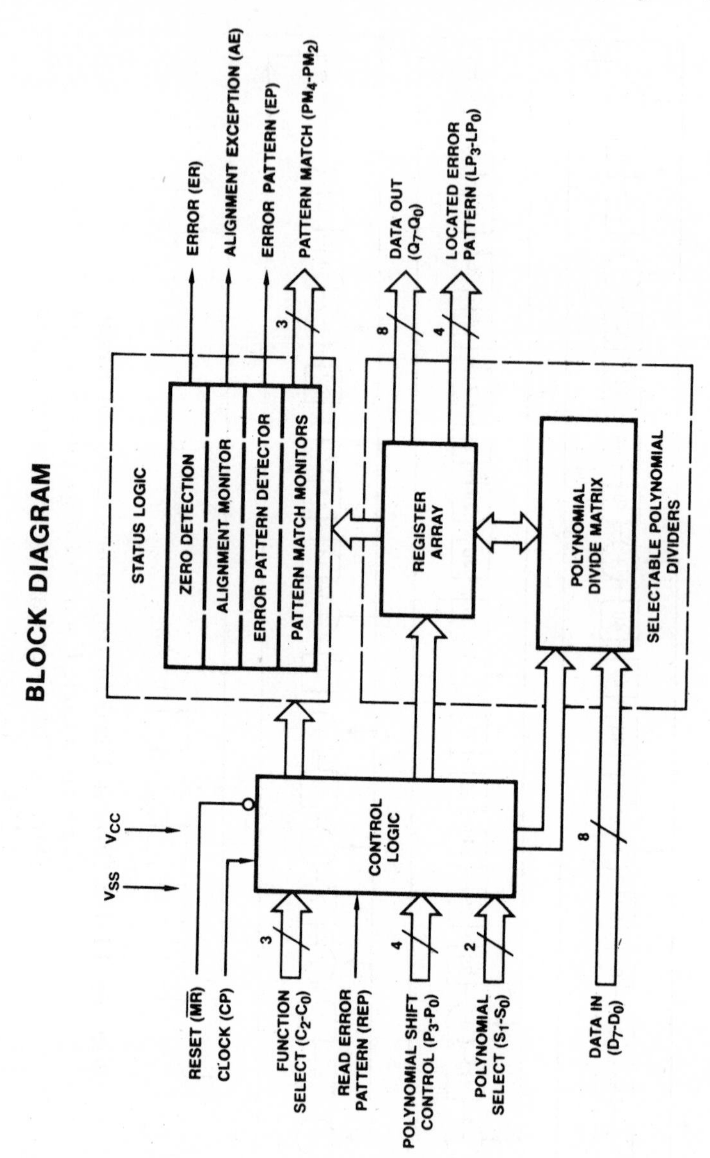

Figure 11-11: Block and Connection Diagrams. Copyright © 1984 Advanced Micro Devices, Inc. Reproduced with permission of copyright owner. All rights reserved.

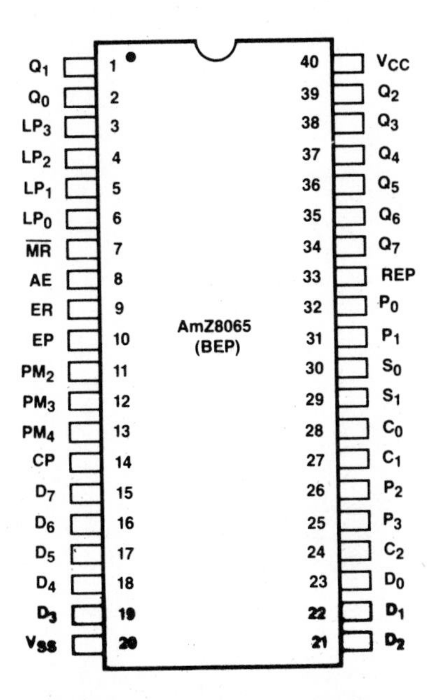

Figure 11-11 (continued)

12 Information on T^2L Logic Chips

The T^2L logic chips in the 74XX series are probably the most popular chips available. They are inexpensive and readily available. Almost all logic functions needed can be found in some combination of the 74XX series of chips.

This chapter is primarily a checklist of the more popular devices. It gives pin-out data as well as other important information such as average current drain and average propagation delay.

The following is a list of the devices covered in this chapter. A short description of each device is given.

DEVICE		DESCRIPTION
7400	NAND gate	quad 2-input
7401	NAND gate	quad 2-input open collector
7402	NOR gate	quad 2-input
7403	NAND gate	quad 2-input open collector
7404	inverter	hex
7405	inverter	hex open collector
7406	driver/inverter	hex open collector high voltage (30 volts)
7407	driver	hex open collector (30 volts) noninverting

DEVICE		DESCRIPTION
7408	AND gate	quad 2-input
7409	AND gate	quad 2-input open collector
7410	NAND gate	triple 3-input
7411	AND gate	triple 3-input
7413	Schmitt trigger	dual NAND gated
7414	Schmitt trigger	hex inverting
7420	NAND gate	dual 4-input
7421	AND gate	dual 4-input
7430	NAND gate	8-input
7432	OR gate	quad 2-input
7437	NAND/buffer	quad 2-input
7440	NAND/buffer	dual 4-input
7441	decoder/driver	BCD-to-decimal
7442	decoder	BCD-to-decimal
7443	decoder	excess 3-to-decimal
7447	decoder/driver	BCD-to-seven segment
7470	flip-flop	JK type
7472	flip-flop	JK master slave
7473	flip-flop	dual JK
7474	flip-flop	dual D-type
7476	flip-flop	dual JK
7477	memory	quadruple latch
7483	full adder	four bit
7486	exclusive OR	quad 2-input
7488	ROM	256 bit
7489	RAM	64 bit
7490	counter	decade
7491	shift register	8-bit
7492	counter	base 12
7493	counter	4-bit binary
7495	shift register	4-bit right & left
74107	flip-flop	dual JK
74121	multivibrator	monostable

DEVICE		DESCRIPTION
74141	decoder/driver	BCD-to-decimal
74150	data selector	1 of 16
74151	data selector	1 of 8
74154	decoder/ demultiplexer	4 line to 16 line
74160	counter	synchronous 4-bit
74164	shift register	8-bit, serial in, parallel out
74165	shift register	8-bit, parallel in, serial out
74174	flip-flop	hex D-type flip-flop
74175	flip-flop	quad D-type
74180	parity	8-bit parity generator checker
74181	arithmetic logic	CPU
74190/91	counter	
74192/93	counter	
74195	shift register	4-bit universal

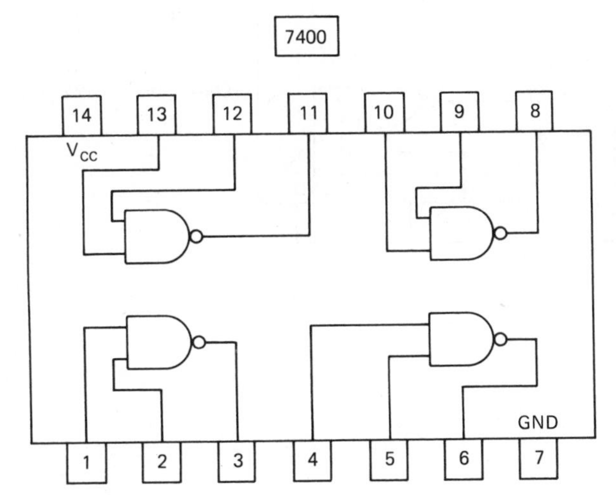

Quadruple 2-Input Positive NAND Gate

Propagation ≈ 10 ns

Current Drain ≈ 12 mA

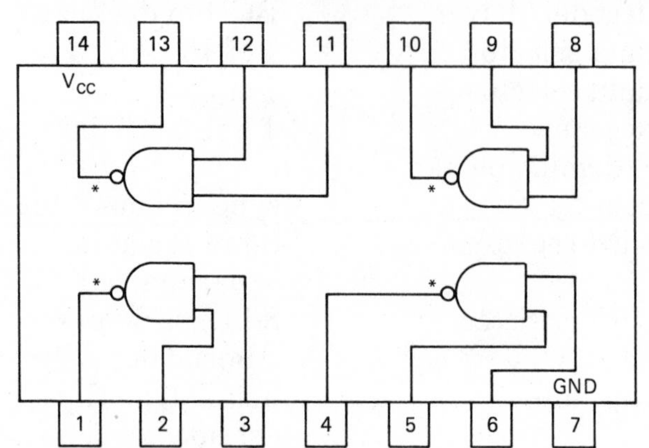

Quadruple 2-Input Positive NAND Gate
(Open Collector)

Current Drain ≈ 8 mA
*Collectors Open

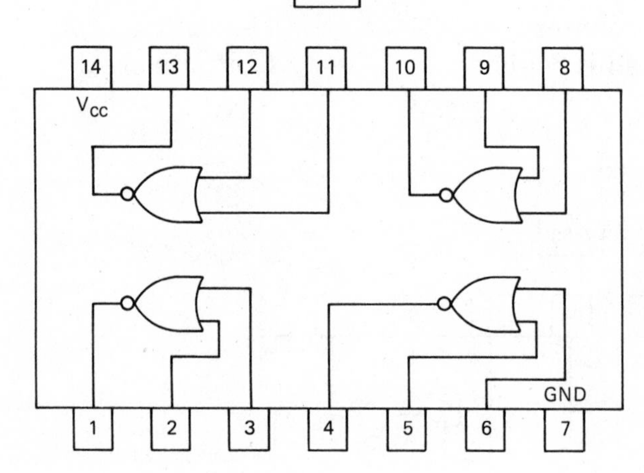

Quadruple 2-Input Positive NOR Gate

Propagation ≈ 10 ns
Current Drain ≈ 12 mA

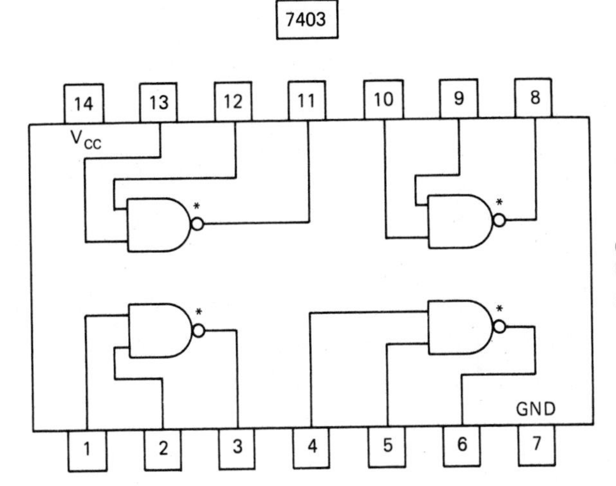

Quadruple 2-Input Positive NAND Gate (Open Collector)

Current Drain ≈ 8 mA

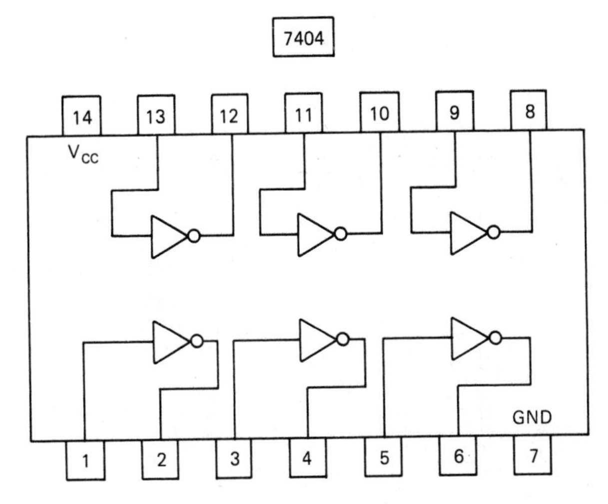

Hex Inverter

Propagation ≈ 10 ns

Current Drain ≈ 8 mA

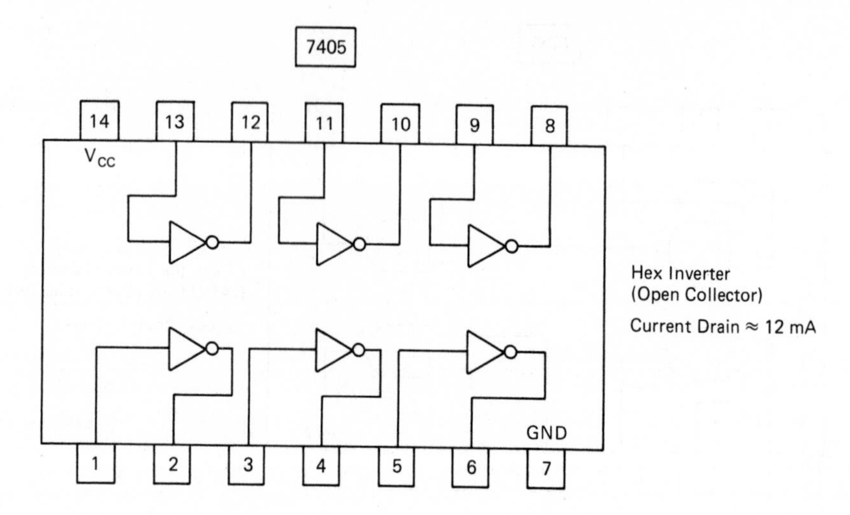

Hex Inverter
(Open Collector)

Current Drain ≈ 12 mA

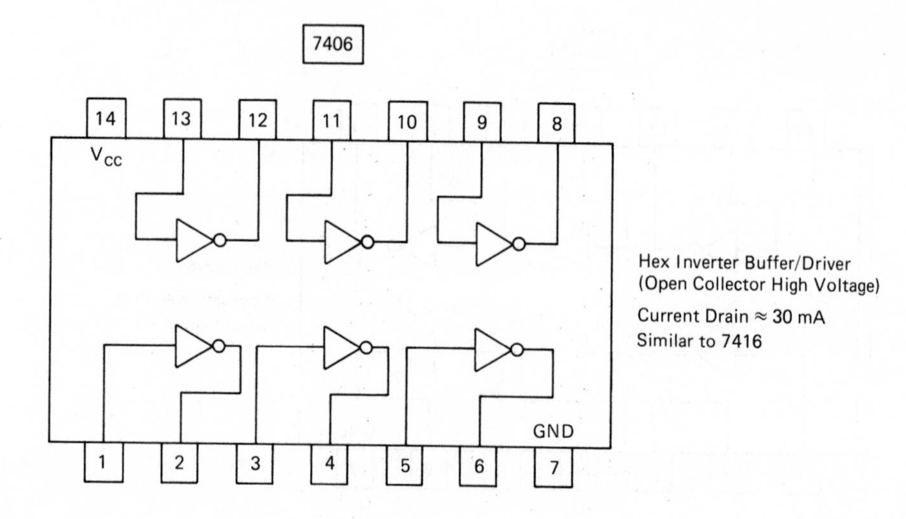

Hex Inverter Buffer/Driver
(Open Collector High Voltage)

Current Drain ≈ 30 mA
Similar to 7416

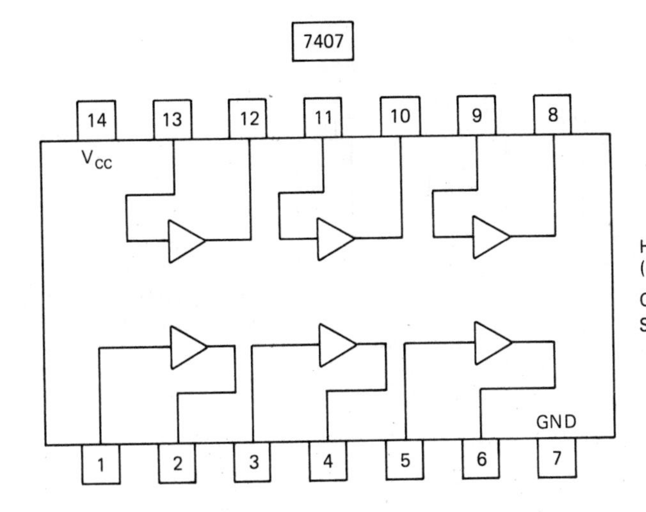

Hex Buffer/Driver
(Open Collector High Voltage)

Current Drain ≈ 25 mA
Similar to 7417

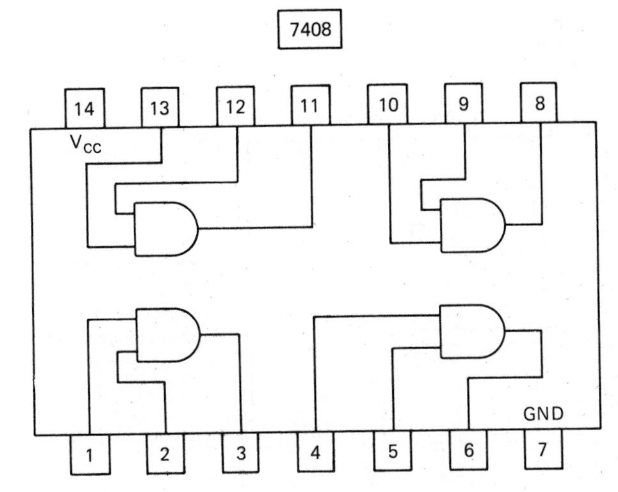

Quadruple 2-Input Positive AND Gate

Propagation ≈ 15 ns
Current Drain ≈ 16 mA

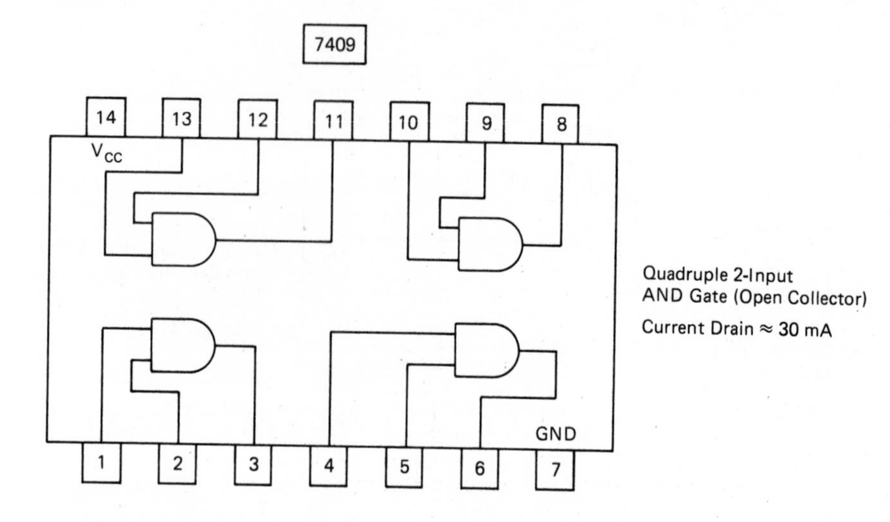
7409
14
13
12
11
10
9
8
V_{CC}
GND
1
2
3
4
5
6
7
Quadruple 2-Input
AND Gate (Open Collector)
Current Drain ≈ 30 mA

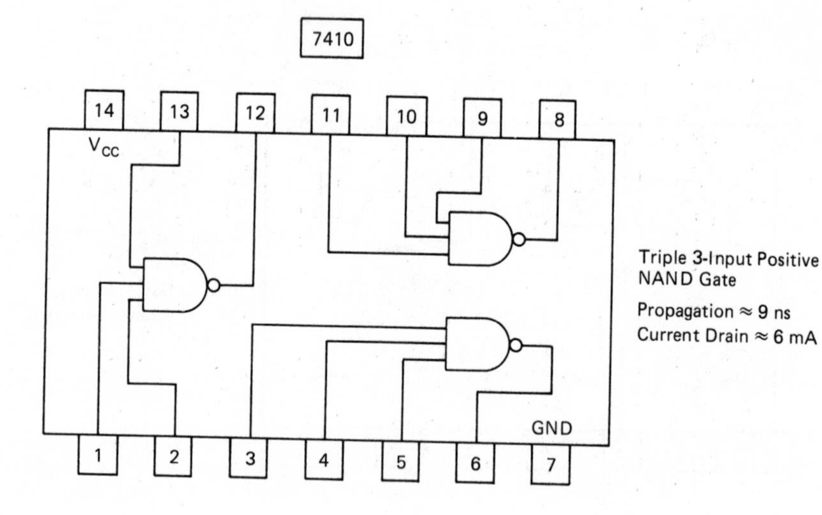
7410
14
13
12
11
10
9
8
V_{CC}
GND
1
2
3
4
5
6
7
Triple 3-Input Positive
NAND Gate
Propagation ≈ 9 ns
Current Drain ≈ 6 mA

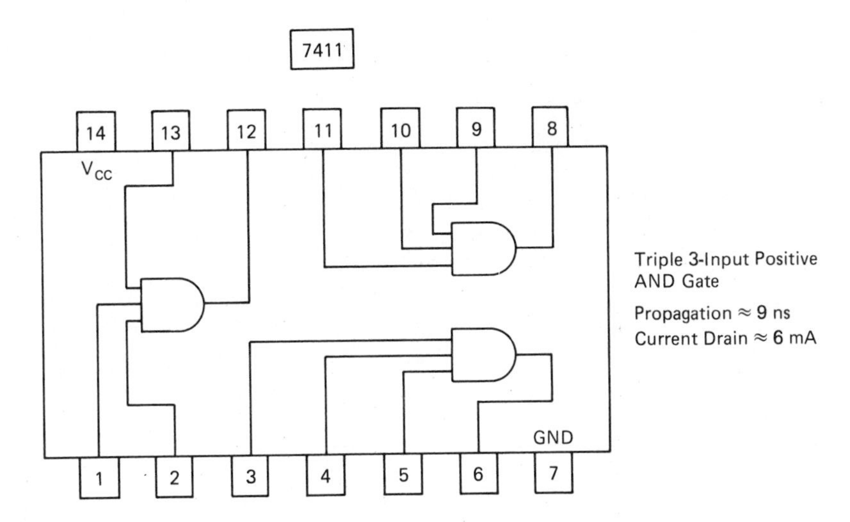
7411
14
13
12
11
10
9
8
V_{CC}
GND
1
2
3
4
5
6
7
Triple 3-Input Positive AND Gate
Propagation ≈ 9 ns
Current Drain ≈ 6 mA

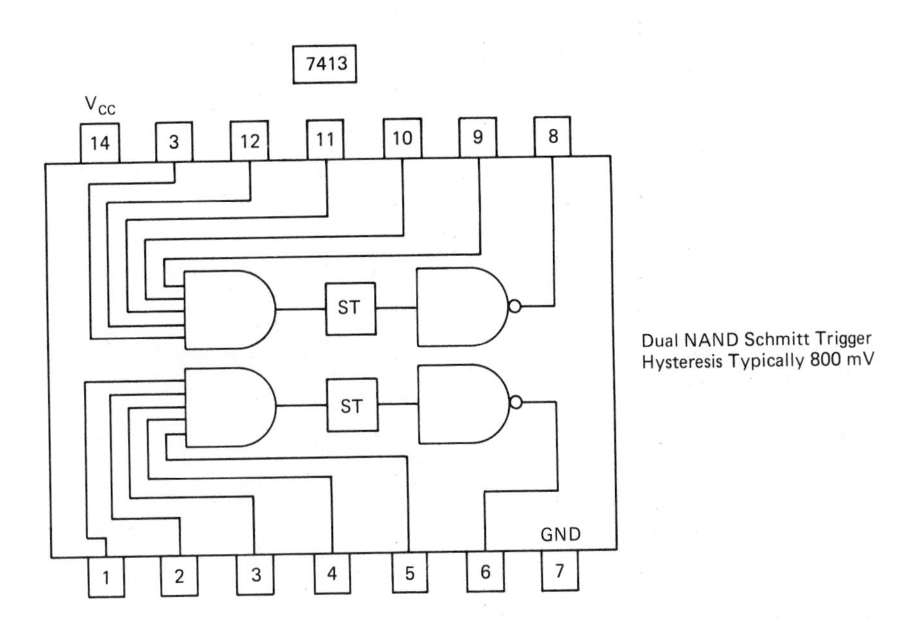
7413
V_{CC}
14
3
12
11
10
9
8
ST
ST
GND
1
2
3
4
5
6
7
Dual NAND Schmitt Trigger
Hysteresis Typically 800 mV

7414

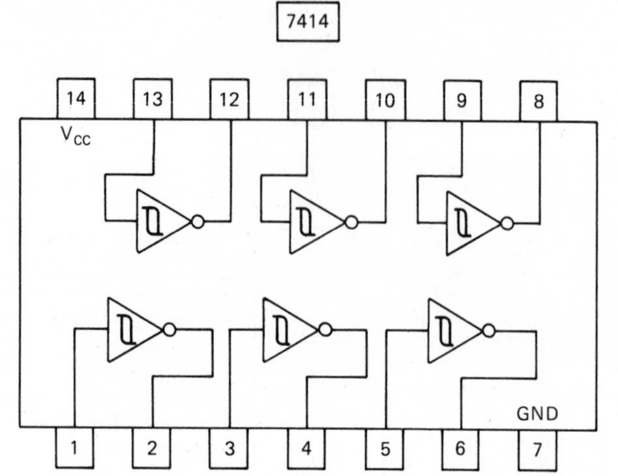

Hex Schmitt Trigger Inverting

Current Drain ≈ 30 mA

Hysteresis Typically 800 mV

7420

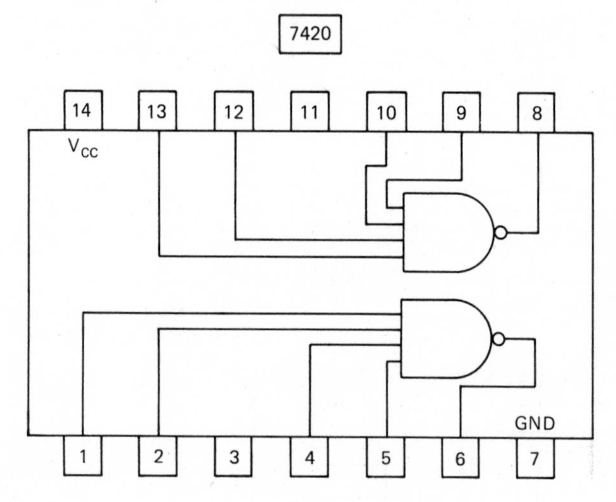

Dual 4-Input Positive NAND Gate

Propagation ≈ 10 ns

Current Drain ≈ 4 mA

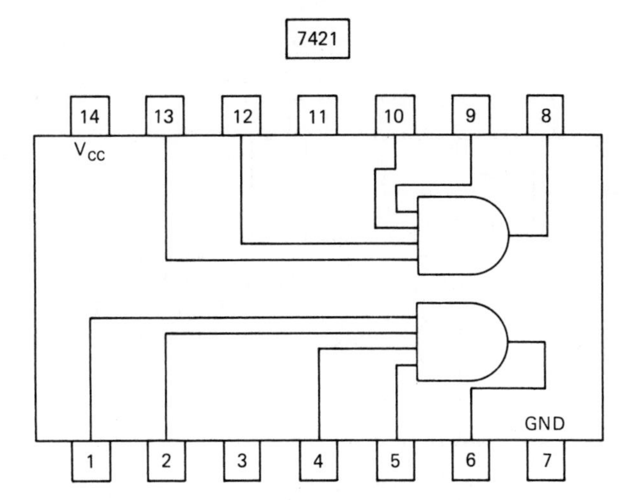

Dual 4-Input Positive AND Gate

Propagation ≈ 10 ns
Current Drain ≈ 5 mA

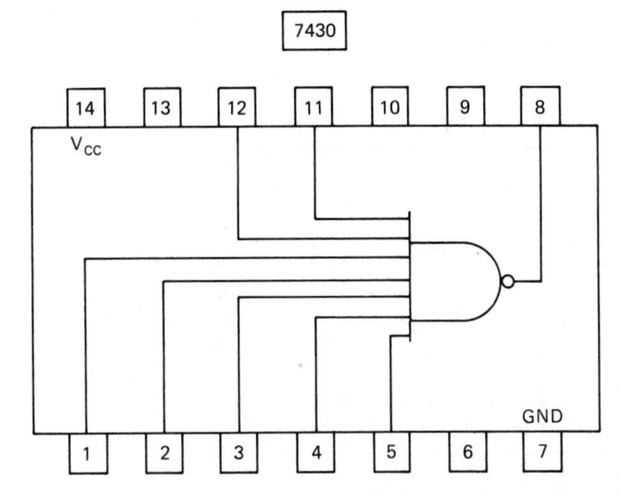

8-Input Positive NAND Gate

Propagation ≈ 10 ns
Current Drain ≈ 2 mA

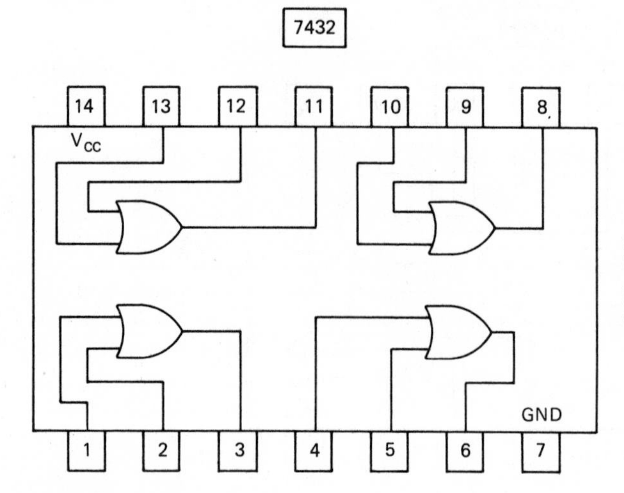

Quad 2-Input Positive OR Gate

Propagation ≈ 12 ns
Current Drain ≈ 19 mA

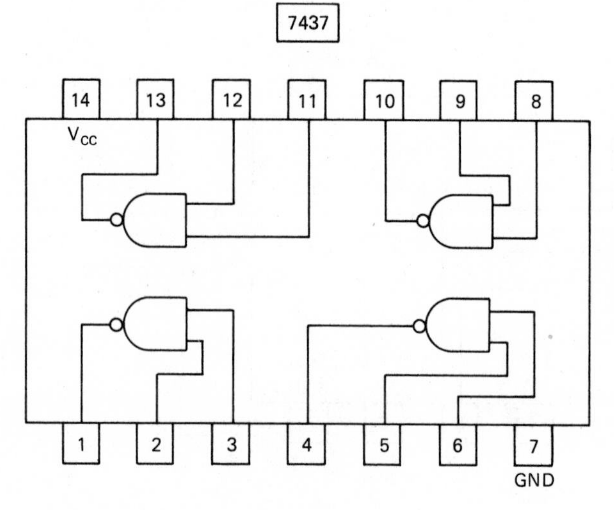

Quadruple 2-Input Positive NAND Buffer

Propagation ≈ 11 ns
Current Drain ≈ 5 mA

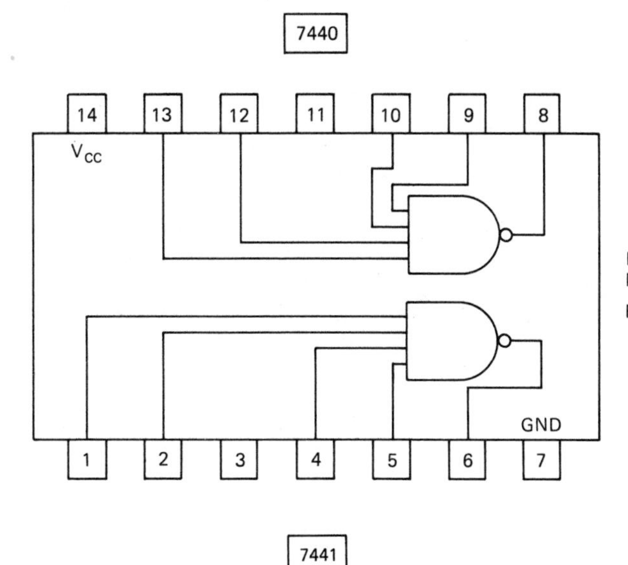

Dual 4-Input Positive NAND Buffer

Propagation ≈ 11 ns

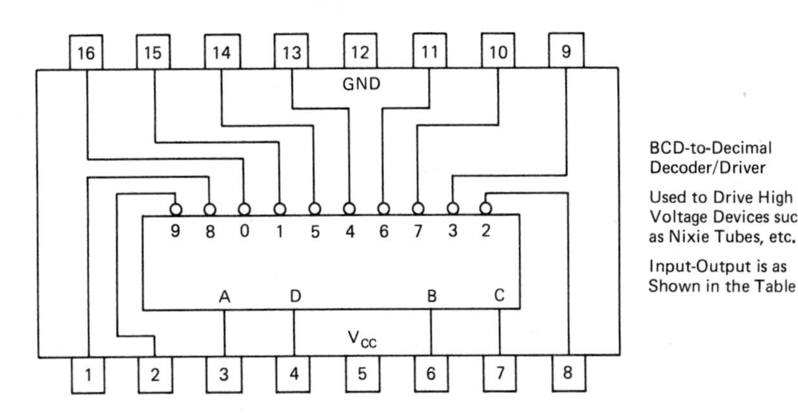

BCD-to-Decimal Decoder/Driver

Used to Drive High Voltage Devices such as Nixie Tubes, etc.

Input-Output is as Shown in the Table

Input				Output On*
D	C	B	A	
0	0	0	0	0
0	0	0	1	1
0	0	1	0	2
0	0	1	1	3
0	1	0	0	4
0	1	0	1	5
0	1	1	0	6
0	1	1	1	7
1	0	0	0	8
1	0	0	1	9

*All Other Outputs Are Off

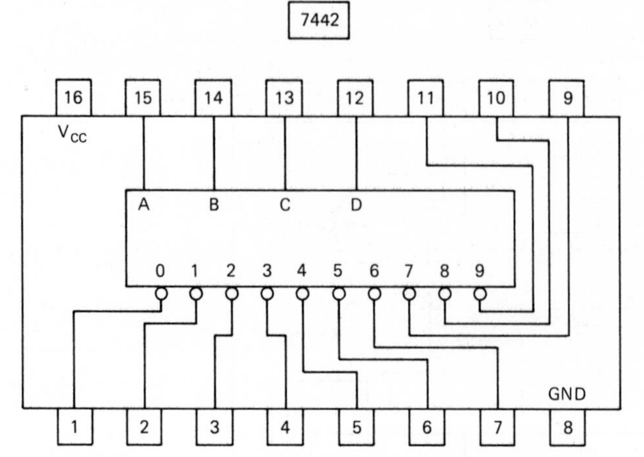

BCD-to-Decimal Decoder
T^2L Output
Propagation ≈ 17 ns
Current Drain ≈ 28 mA

Input				Output Off*
D	C	B	A	
0	0	0	0	0
0	0	0	1	1
0	0	1	0	2
0	0	1	1	3
0	1	0	0	4
0	1	0	1	5
0	1	1	0	6
0	1	1	1	7
1	0	0	0	8
1	0	0	1	9

*All other outputs are on

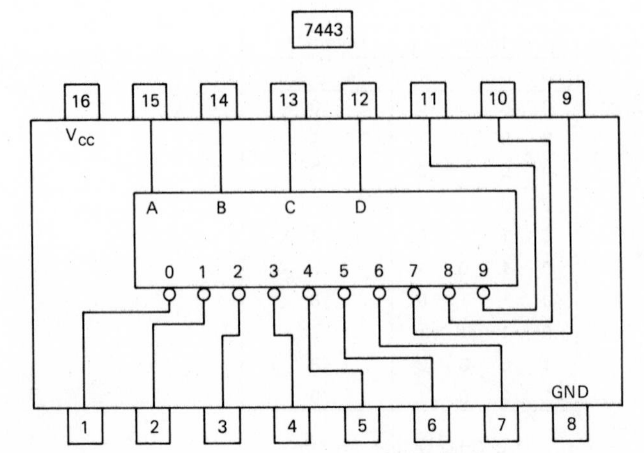

Excess 3-to-Decimal
Decoder

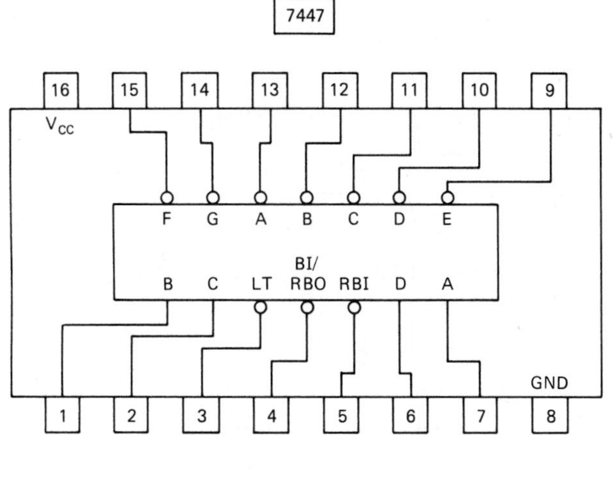

BCD-to-Seven-Segment Decoder/Driver

LT = Lamp Test
BI = Blanking Input
RBO = Ripple Blanking Output
RBI = Ripple Blanking Input

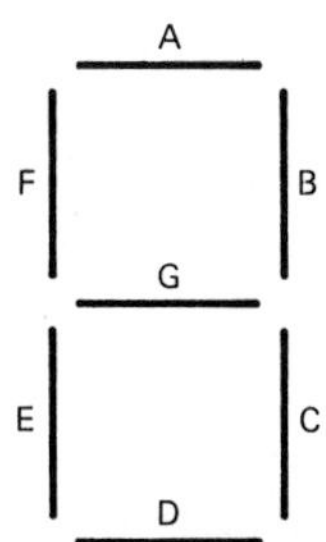

Segment Identification

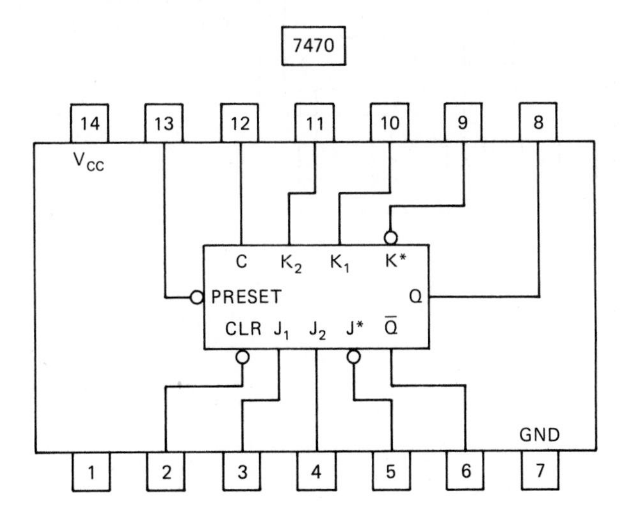

J-K Flip-Flop Positive Edge Triggered

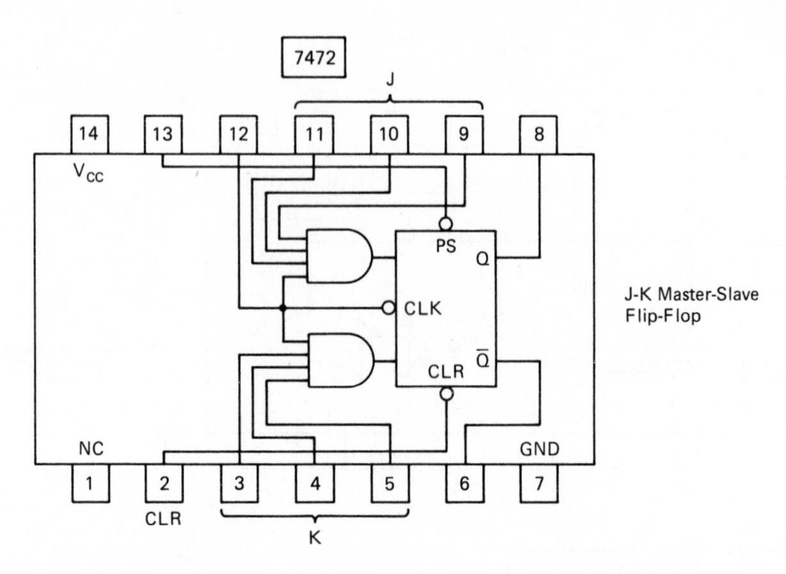
7472
J
14
13
12
11
10
9
8
V_{CC}
PS
Q
CLK
CLR
$\overline{Q}$
NC
GND
1
2
3
4
5
6
7
CLR
K
J-K Master-Slave Flip-Flop

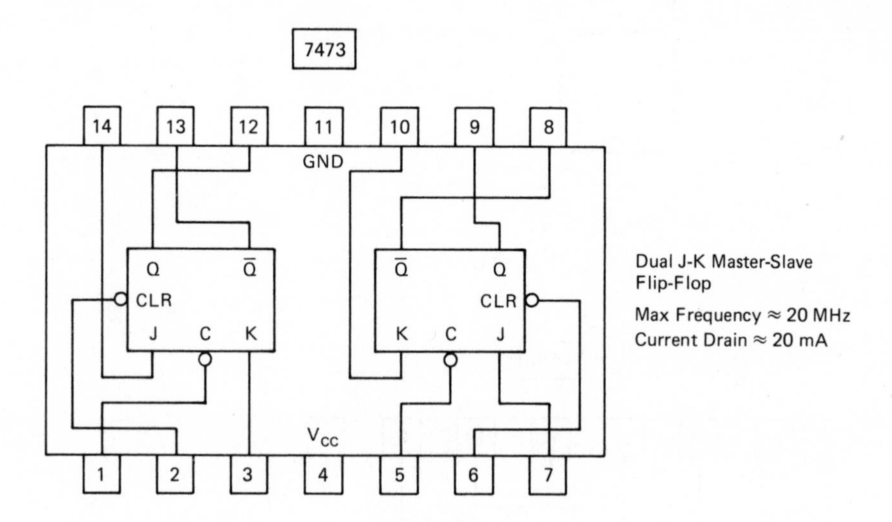
7473
14
13
12
11
10
9
8
GND
Q
$\overline{Q}$
CLR
J
C
K
$\overline{Q}$
Q
CLR
K
C
J
V_{CC}
1
2
3
4
5
6
7
Dual J-K Master-Slave Flip-Flop
Max Frequency ≈ 20 MHz
Current Drain ≈ 20 mA

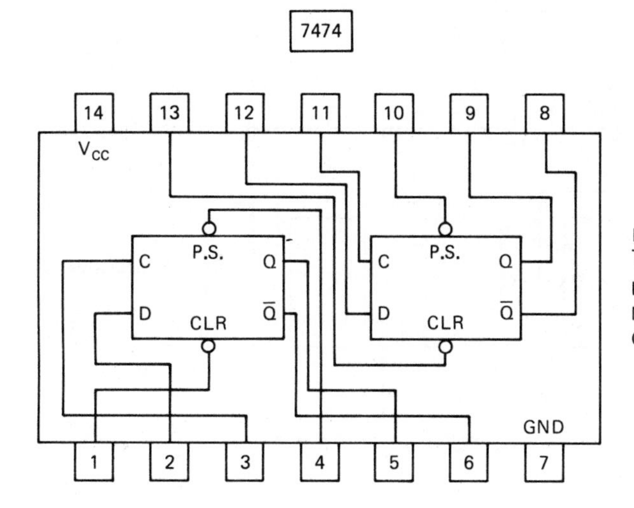

Dual D-Type Edge Triggered Flip-Flop

Positive Edge Triggered
Max Frequency ≈ 25 MHz
Current Drain ≈ 17 mA

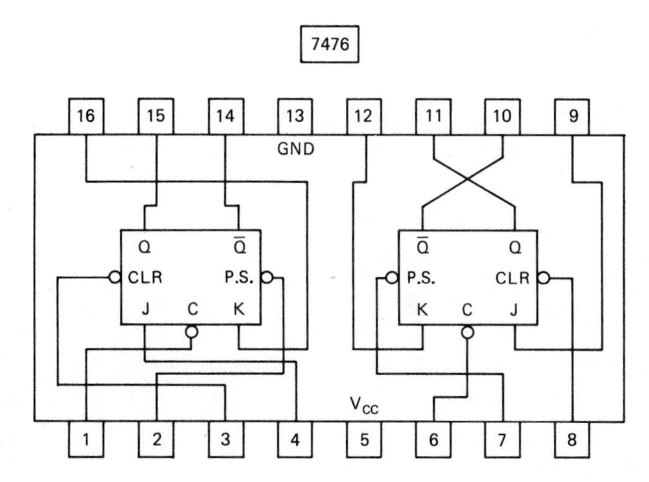

Dual J-K Master-Slave Flip-Flop

Max Frequency ≈ 20 MHz
Current Drain ≈ 20 mA

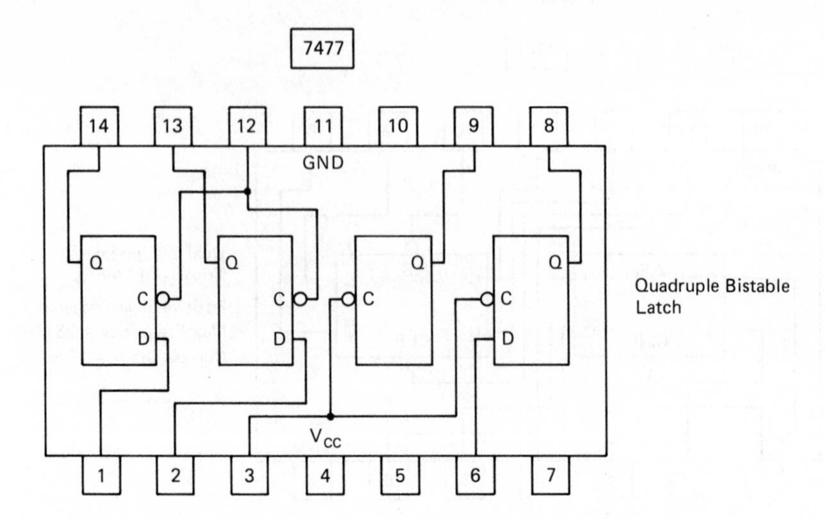
7477
14
13
12
11
10
9
8
GND
Q
C
D
Q
C
D
C
Q
C
Q
D
V_CC
1
2
3
4
5
6
7
Quadruple Bistable
Latch

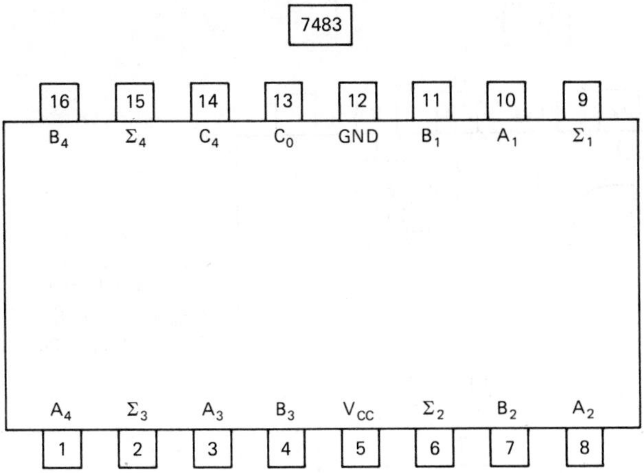

7483

16	15	14	13	12	11	10	9
B_4	Σ_4	C_4	C_0	GND	B_1	A_1	Σ_1
A_4	Σ_3	A_3	B_3	V_{CC}	Σ_2	B_2	A_2
1	2	3	4	5	6	7	8

Full Adder 4-Bit with Look Ahead Carry

Propagation ≈ 12 ns

In				Out					
A_1 / A_3	B_1 / C_3	A_2 / A_4	B_2 / B_4	Σ_1 / Σ_3	Σ_2 / Σ_4	C_2 / C_4	Σ_1 / Σ_3	Σ_2 / Σ_4	C_2 / C_4
0	0	0	0	0	0	0	1	0	0
1	0	0	0	1	0	0	0	1	0
0	1	0	0	1	0	0	0	1	0
1	1	0	0	0	1	0	1	1	0
0	0	1	0	0	1	0	1	1	0
1	0	1	0	1	1	0	0	0	1
0	1	1	0	1	1	0	0	0	1
1	1	1	0	0	0	1	1	0	1
0	0	0	1	0	1	0	1	1	0
1	0	0	1	1	1	0	0	0	1
0	1	0	1	1	1	0	0	0	1
1	1	0	1	0	0	1	1	0	1
0	0	1	1	0	0	1	1	0	1
1	0	1	1	1	0	1	0	1	1
0	1	1	1	1	0	1	0	1	1
1	1	1	1	0	1	1	1	1	1

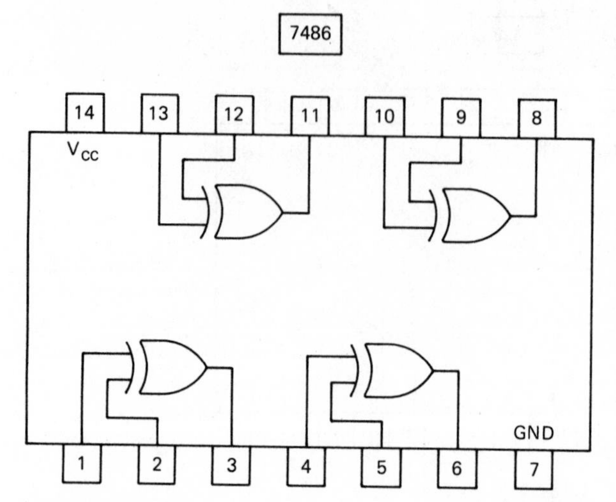

Quadruple 2-Input Exclusive OR Gate

Propagation ≈ 18 ns
Current Drain ≈ 30 mA

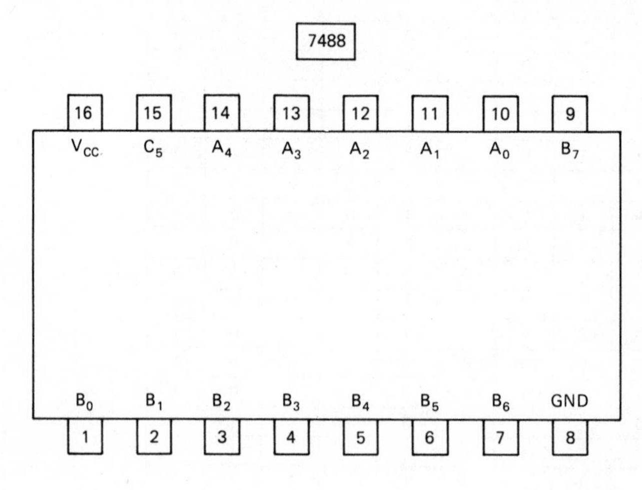

256-Bit Read-Only-Memory

32-X 8-Bit Words
Power 400mW max

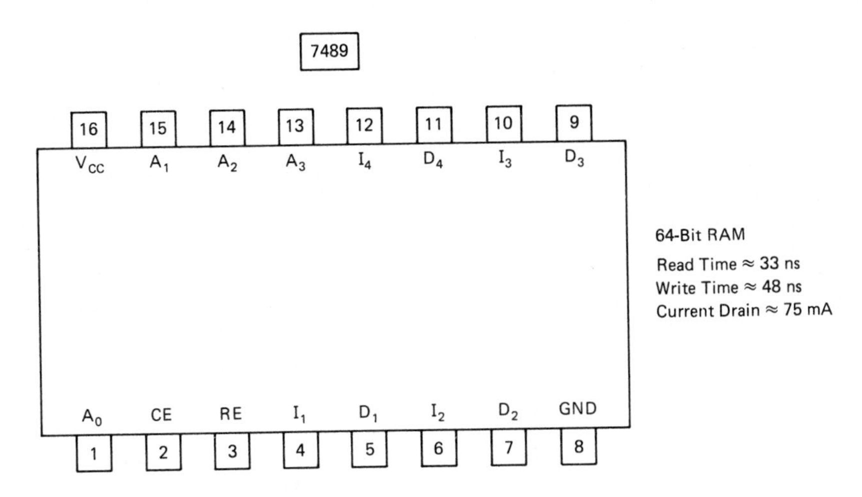

64-Bit RAM

Read Time ≈ 33 ns

Write Time ≈ 48 ns

Current Drain ≈ 75 mA

Decade Counter

Divide-by-Two/
Divide-by-Five

Max. Frequency ≈ 18 MHz

Current Drain ≈ 32 mA

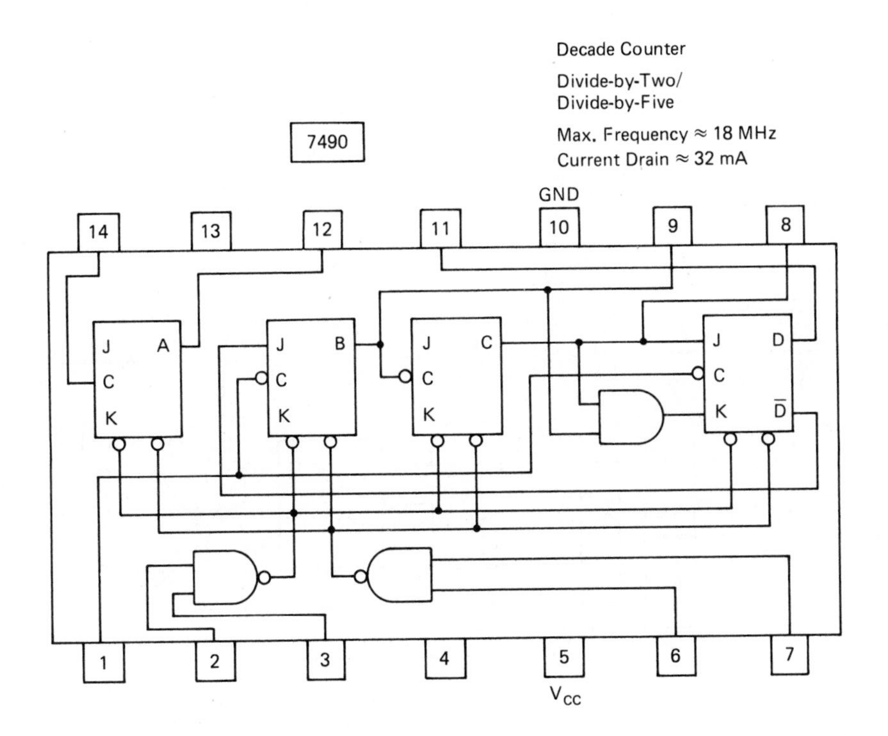

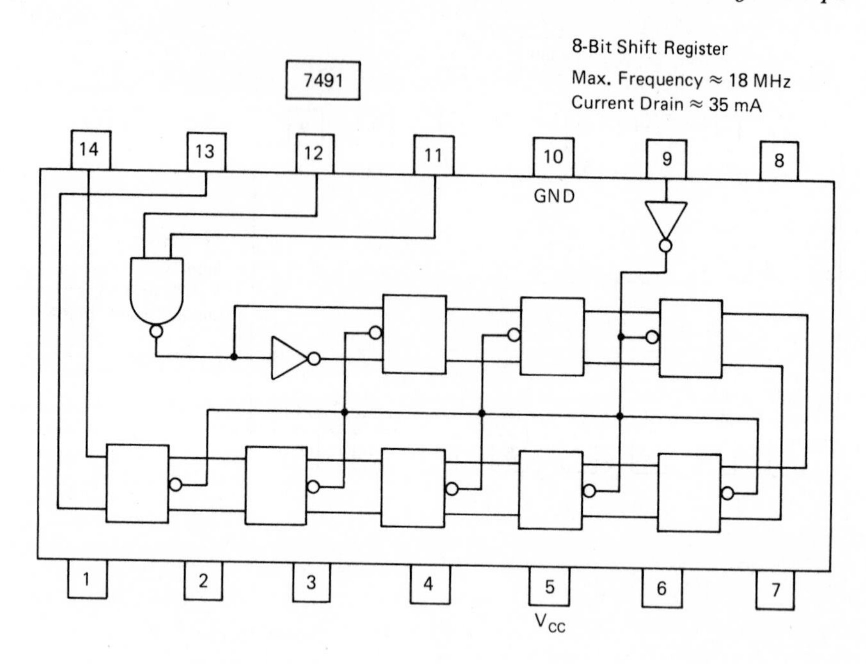
8-Bit Shift Register
Max. Frequency ≈ 18 MHz
Current Drain ≈ 35 mA
7491
14
13
12
11
10
9
8
GND
1
2
3
4
5
6
7
V_{CC}

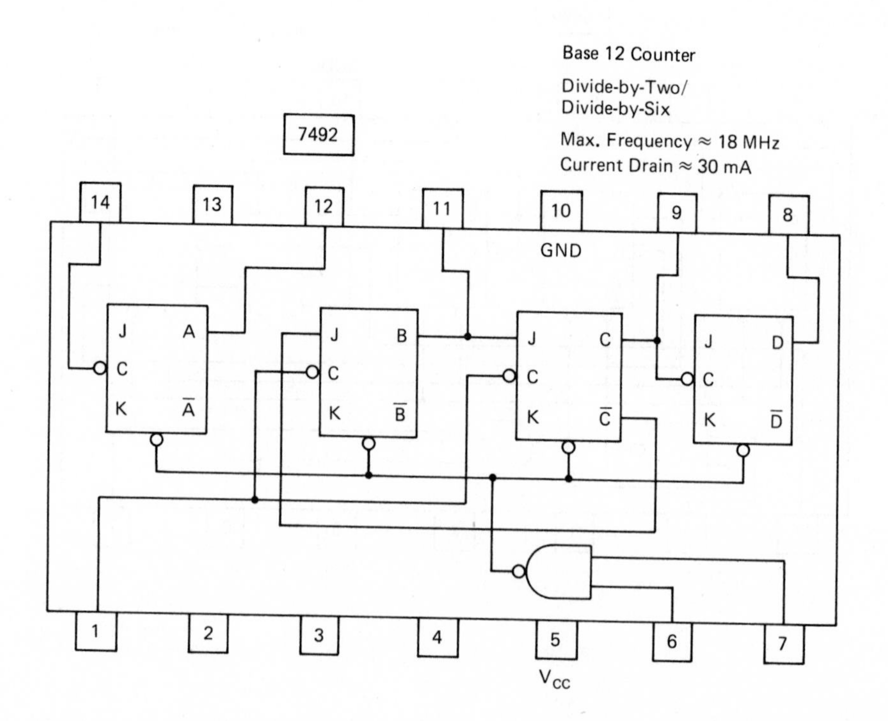
Base 12 Counter
Divide-by-Two/
Divide-by-Six
Max. Frequency ≈ 18 MHz
Current Drain ≈ 30 mA
7492
14
13
12
11
10
9
8
GND
J
C
K
A
$\overline{A}$
B
$\overline{B}$
C
$\overline{C}$
D
$\overline{D}$
1
2
3
4
5
6
7
V_{CC}

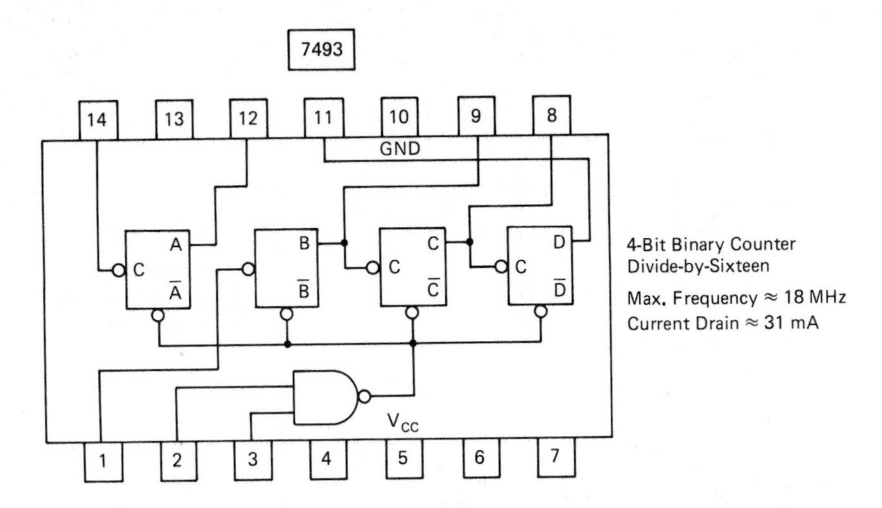

4-Bit Binary Counter
Divide-by-Sixteen

Max. Frequency ≈ 18 MHz
Current Drain ≈ 31 mA

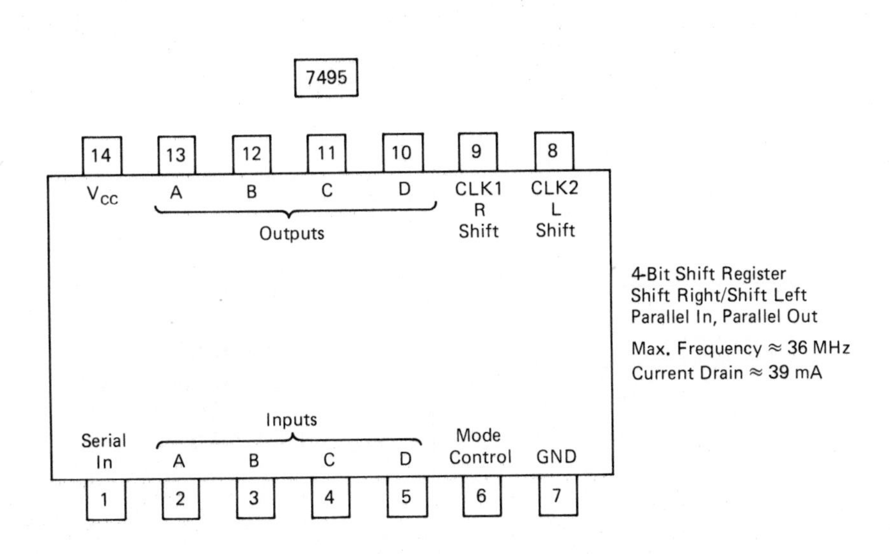

4-Bit Shift Register
Shift Right/Shift Left
Parallel In, Parallel Out

Max. Frequency ≈ 36 MHz
Current Drain ≈ 39 mA

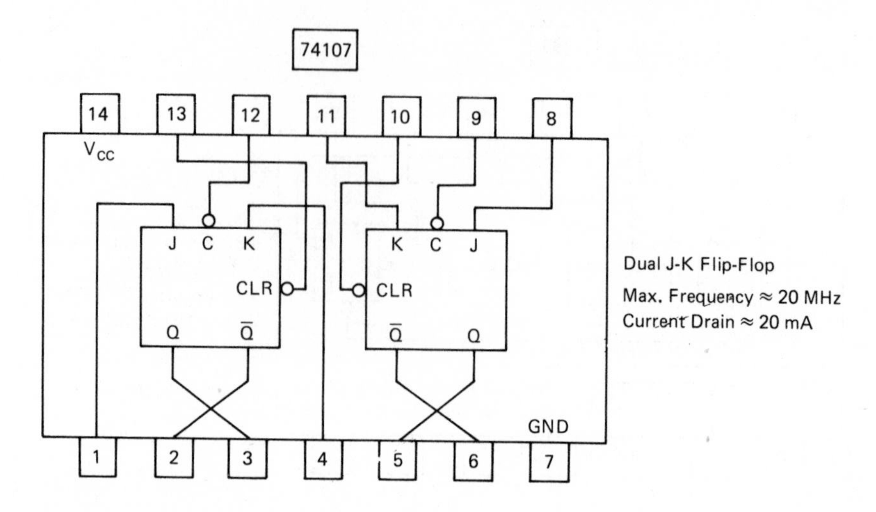

Dual J-K Flip-Flop

Max. Frequency ≈ 20 MHz

Current Drain ≈ 20 mA

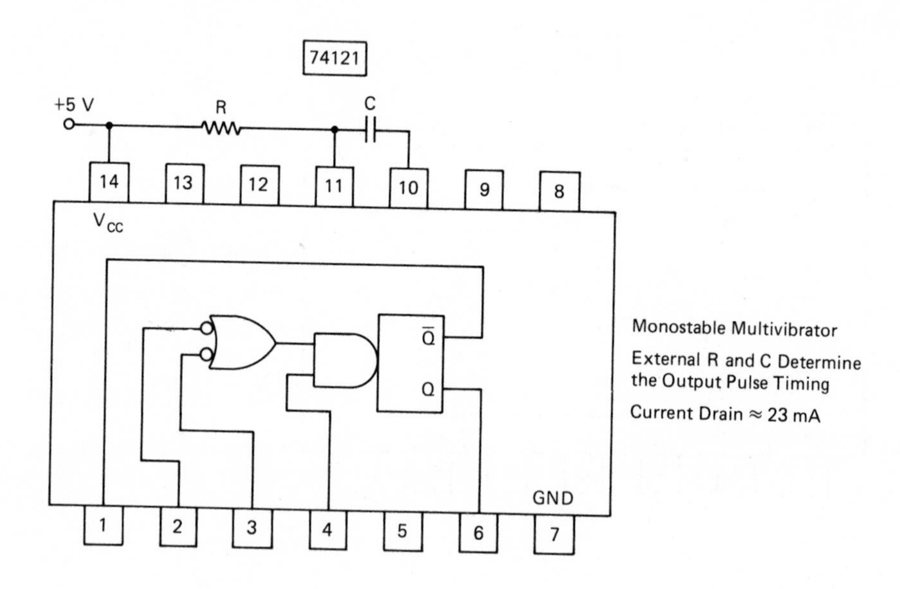

Monostable Multivibrator

External R and C Determine the Output Pulse Timing

Current Drain ≈ 23 mA

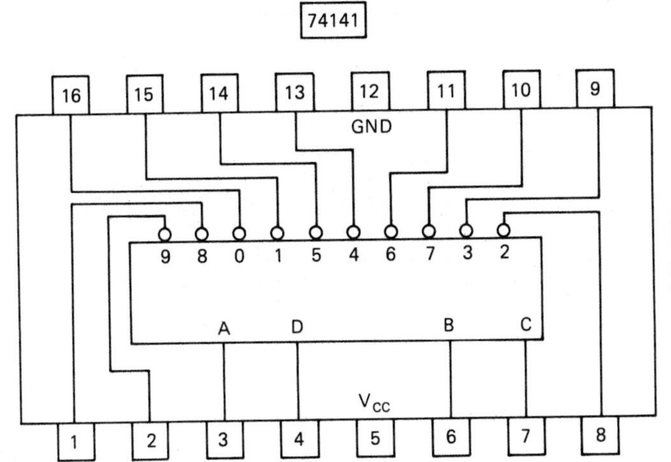

BCD-to-Decimal Decoder/Driver

Nixie Tube Driver
7 mA, 60 volt

D	C	B	A	Output On
0	0	0	0	0
0	0	0	1	1
0	0	1	0	2
0	0	1	1	3
0	1	0	0	4
0	1	0	1	5
0	1	1	0	6
0	1	1	1	7
1	0	0	0	8
1	0	0	1	9

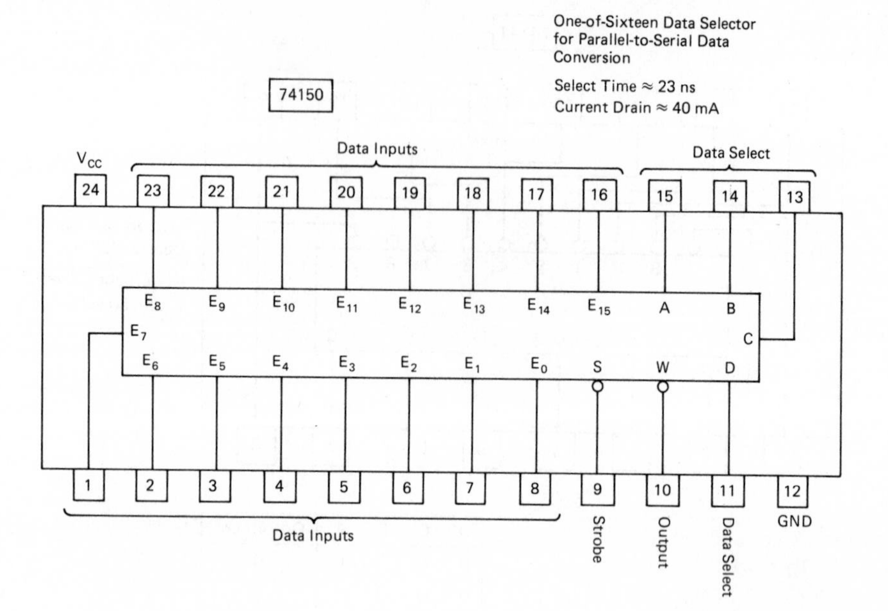

One-of-Sixteen Data Selector for Parallel-to-Serial Data Conversion
Select Time ≈ 23 ns
Current Drain ≈ 40 mA
74150
Vcc
Data Inputs
Data Select
24
23
22
21
20
19
18
17
16
15
14
13
E8
E9
E10
E11
E12
E13
E14
E15
A
B
E7
C
E6
E5
E4
E3
E2
E1
E0
S
W
D
1
2
3
4
5
6
7
8
9
10
11
12
Data Inputs
Strobe
Output
Data Select
GND

74150 Continued

Inputs																					Out
D	C	B	A	Strobe	E_0	E_1	E_2	E_3	E_4	E_5	E_6	E_7	E_8	E_9	E_{10}	E_{11}	E_{12}	E_{13}	E_{14}	E_{15}	W
0	0	0	0	0	0																1
0	0	0	0	0	1																0
0	0	0	1	0		0															1
0	0	0	1	0		1															0
0	0	1	0	0			0														1
0	0	1	0	0			1														0
0	0	1	1	0				0													1
0	0	1	1	0				1													0
0	1	0	0	0					0												1
0	1	0	0	0					1												0
0	1	0	1	0						0											1
0	1	0	1	0						1											0
0	1	1	0	0							0										1
0	1	1	0	0							1										0
0	1	1	1	0								0									1
0	1	1	1	0								1									0
1	0	0	0	0									0								1
1	0	0	0	0									1								0
1	0	0	1	0										0							1
1	0	0	1	0										1							0
1	0	1	0	0											0						1
1	0	1	0	0											1						0
1	0	1	1	0												0					1
1	0	1	1	0												1					0
1	1	0	0	0													0				1
1	1	0	0	0													1				0
1	1	0	1	0														0			1
1	1	0	1	0														1			0
1	1	1	0	0															0		1
1	1	1	0	0															1		0
1	1	1	1	0																0	1
1	1	1	1	0																1	0

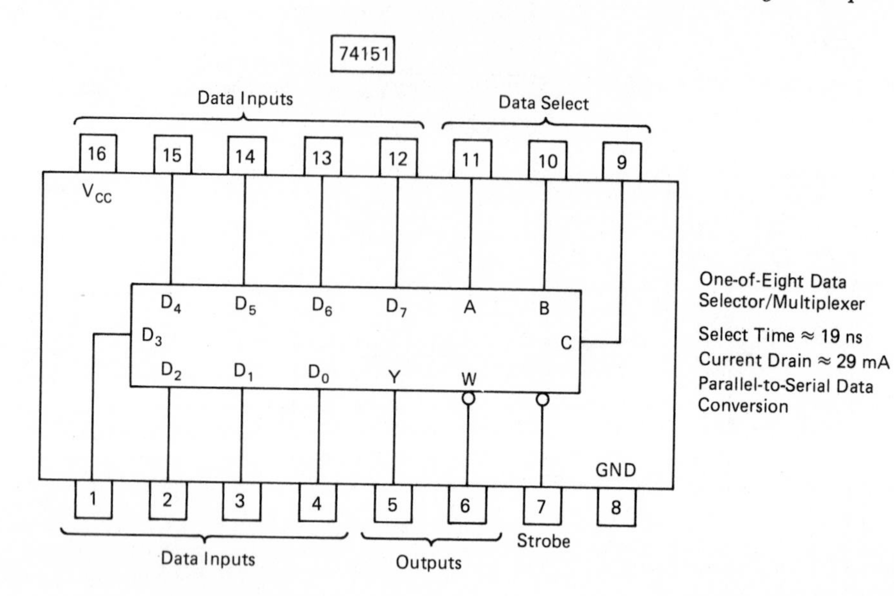

One-of-Eight Data Selector/Multiplexer

Select Time ≈ 19 ns

Current Drain ≈ 29 mA

Parallel-to-Serial Data Conversion

Inputs												Out	
C	B	A	Strobe	D_0	D_1	D_2	D_3	D_4	D_5	D_6	D_7	Y	W
0	0	0	0	0								0	1
0	0	0	0	1								1	0
0	0	1	0		0							0	1
0	0	1	0		1							1	0
0	1	0	0			0						0	1
0	1	0	0			1						1	0
0	1	1	0				0					0	1
0	1	1	0				1					1	0
1	0	0	0					0				0	1
1	0	0	0					1				1	0
1	0	1	0						0			0	1
1	0	1	0						1			1	0
1	1	0	0							0		0	1
1	1	0	0							1		1	0
1	1	1	0								0	0	1
1	1	1	0								1	1	0

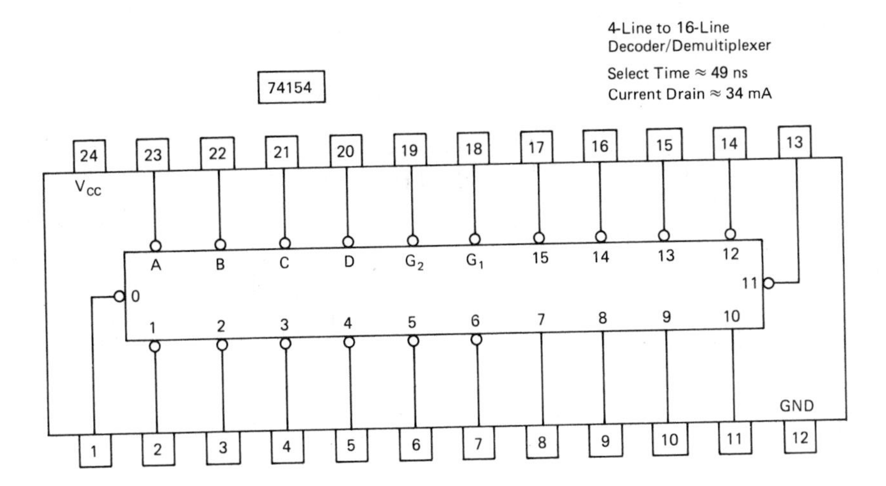

Inputs						Outputs															
G_1	G_2	D	C	B	A	0	1	2	3	4	5	6	7	8	9	10	11	12	13	14	15
0	0	0	0	0	0	0															
0	0	0	0	0	1		0														
0	0	0	0	1	0			0													
0	0	0	0	1	1				0												
0	0	0	1	0	0					0											
0	0	0	1	0	1						0										
0	0	0	1	1	0							0									
0	0	0	1	1	1								0								
0	0	1	0	0	0									0							
0	0	1	0	0	1										0						
0	0	1	0	1	0											0					
0	0	1	0	1	1												0				
0	0	1	1	0	0													0			
0	0	1	1	0	1														0		
0	0	1	1	1	0															0	
0	0	1	1	1	1																0
0	1	X	X	X	X																
1	0	X	X	X	X																
1	1	X	X	X	X																

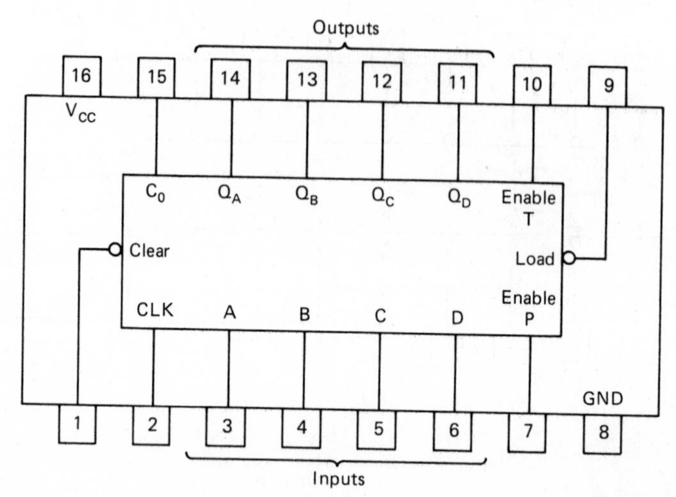

Synchronous 4-Bit/Decade Counter

Max. Frequency ≈ 25 MHz
Current Drain ≈ 34 mA

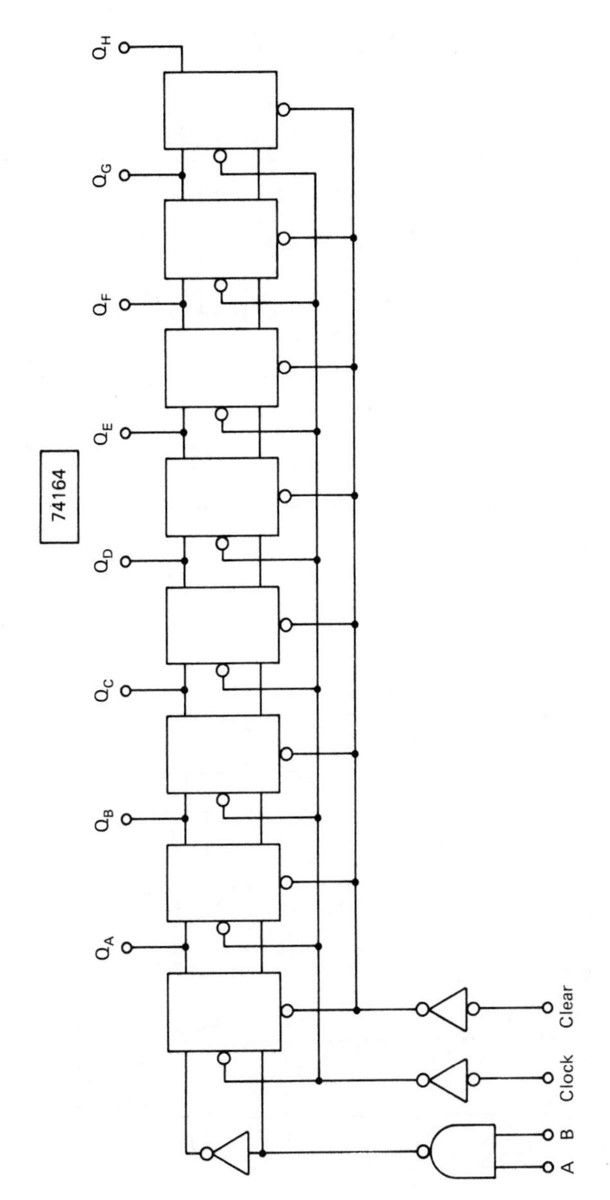
74164
Q_H
Q_G
Q_F
Q_E
Q_D
Q_C
Q_B
Q_A
Clear
Clock
A B

74164

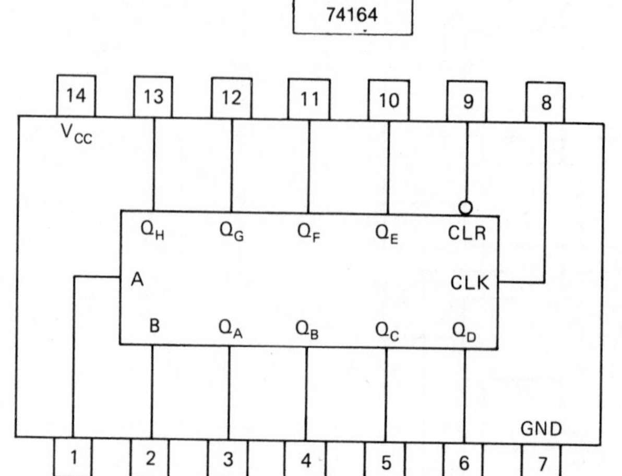

8-Bit Shift Register
Serial In, Parallel Out

Max. Frequency ≈ 36 MHz
Current Drain ≈ 37 mA

74165

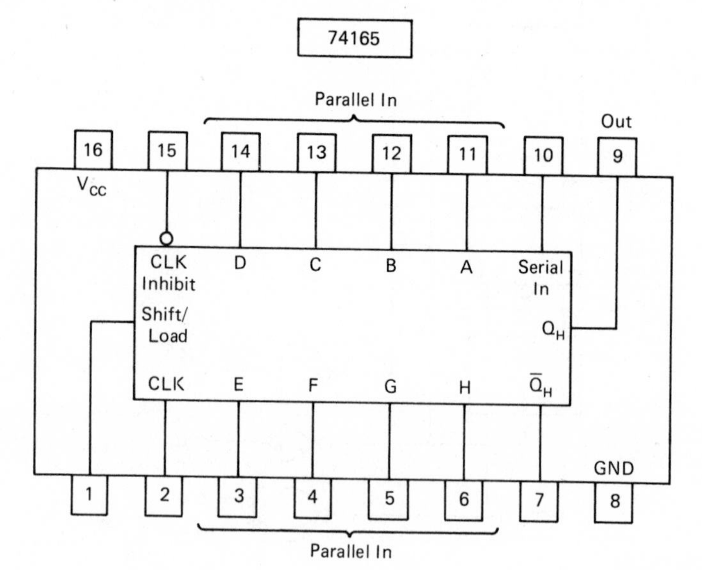

8-Bit Shift Register
Parallel In, Serial Out

Max. Frequency ≈ 26 MHz
Current Drain ≈ 42 mA

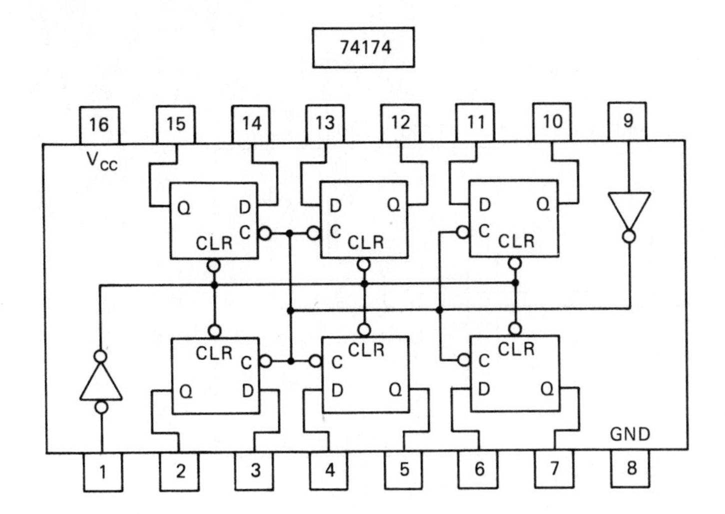

Flip-Flop, Hex D-Type
Positive Edge Triggered
Propagation Delay 23 ns

Max. Frequency ≈ 35 MHz
Current Drain ≈ 45 mA

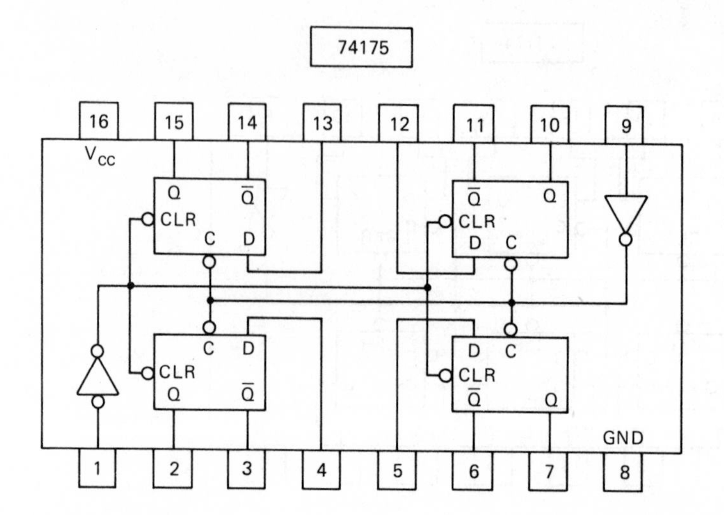

Quadruple D-Type
Flip-Flops Positive
Edge Triggered

Max. Frequency ≈ 35 MHz
Current Drain ≈ 30 mA

Input	Output
T_N	$T_N + 1$
D	Q
H	H
L	L

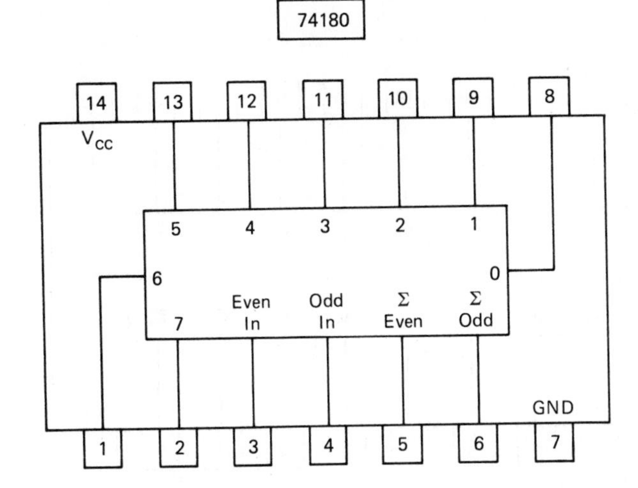

8-Bit, Even/Odd Parity Generator/Checker

Propagation ≈ 45 ns

Current Drain ≈ 34 mA

Input at 0→7	Inputs: Even	Inputs: Odd	Σ Even	Σ Odd
Even	1	0	1	0
Odd	1	0	0	1
Even	0	1	0	1
Odd	0	1	1	0
X	1	1	0	0
X	0	0	1	1

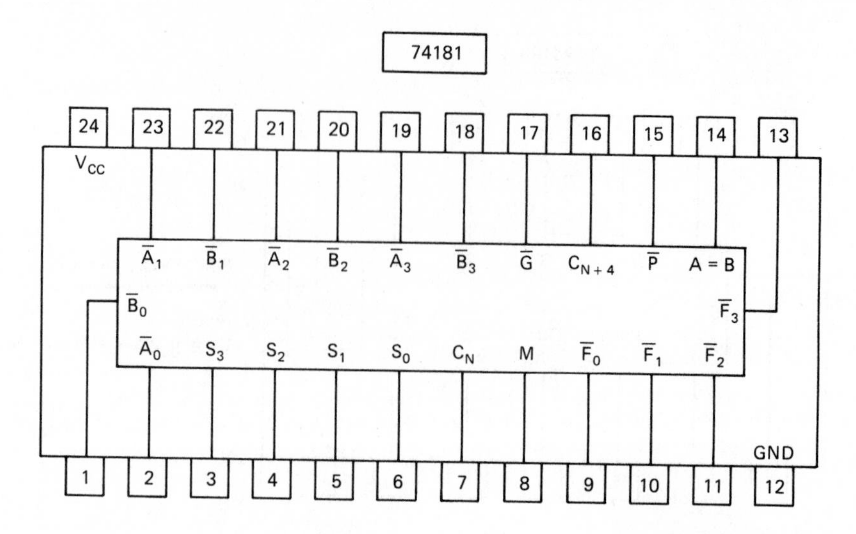

Arithmetic Logic

Operating Time ≈ 35 ns

Current Drain ≈ 94 mA

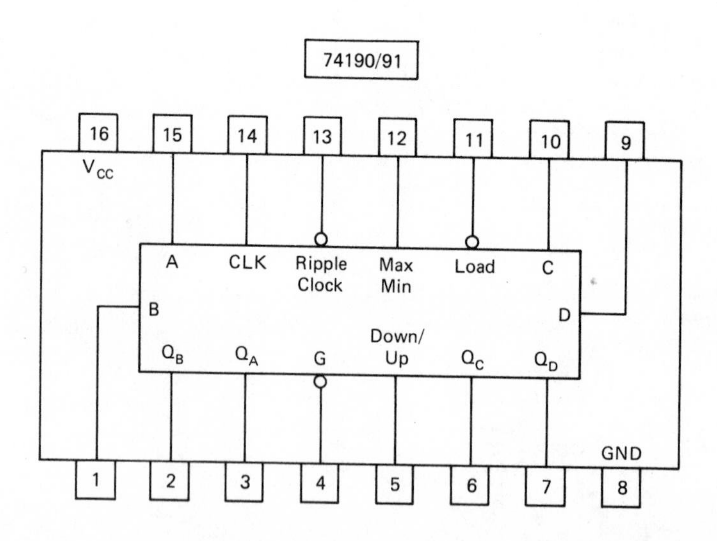

Up/Down Counter Synchronous

Max. Frequency ≈ 20 MHz

Current Drain ≈ 65 mA

xx190 – BCD

xx191 – Binary

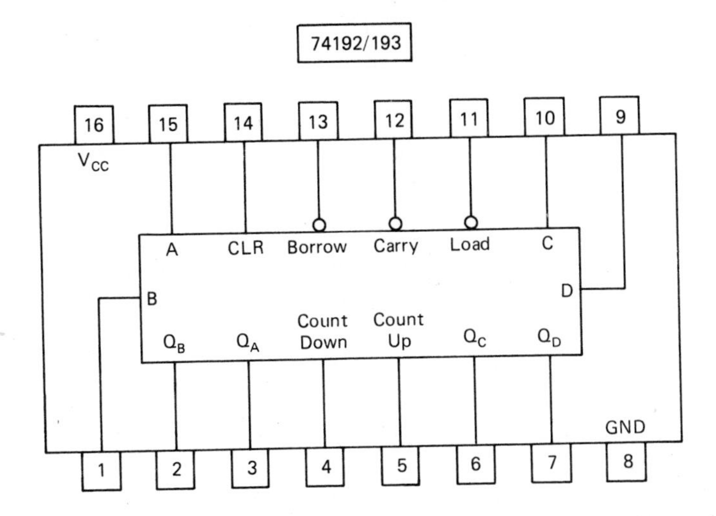

Up/Down Counters, Synchronous with Preset Inputs

xx192 – B.C.D.

xx193 – Binary

Max. Frequency ≈ 32 MHz

Current Drain ≈ 65 mA

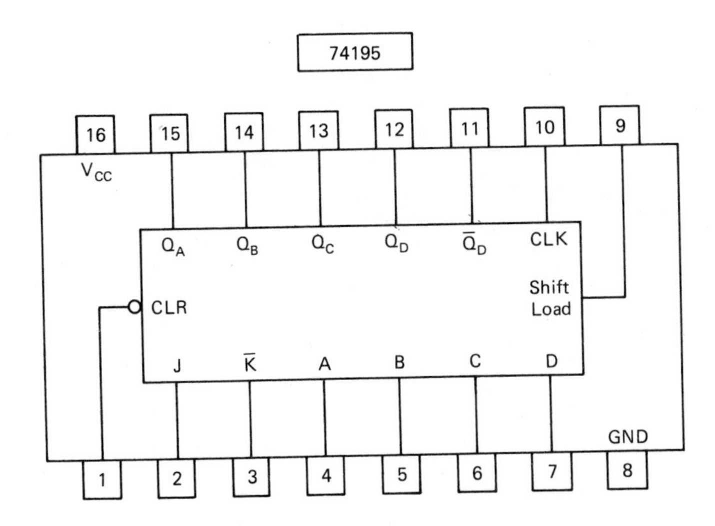

Shift Register, 4-Bit Universal

Current Drain ≈ 20 mA

INDEX